How to Choose and Use Audio Interfaces

MUSICIAN'S GUIDE TO
homerecording

How to Choose and Use
Audio Interfaces

Craig Anderton

Hal Leonard Books
An Imprint of Hal Leonard LLC

Published in 2018 by Hal Leonard Books
An Imprint of Hal Leonard LLC
7777 West Bluemound Road
Milwaukee, WI 53213

Trade Book Division Editorial Offices
33 Plymouth St., Montclair, NJ 07042

The following photos are provided courtesy of Sweetwater.com: Figures 2.6, 5.6, and 6.1.

Printed in the United States of America

Book design by NextPoint Training, Inc.

Library of Congress Cataloging-in-Publication Data

Names: Anderton, Craig, author.
Title: How to choose and use audio interfaces / Craig Anderton.
Description: Milwaukee, WI : Hal Leonard Books, an imprint of Hal Leonard LLC, 2018. | Series: Musician's guide to home recording | Includes bibliographical references.
Identifiers: LCCN 2018022623 | ISBN 9781540024916
Subjects: LCSH: Computer sound processing. | Sound--Recording and reproducing--Digital techniques. | Music--Data processing. | Software synthesizers.
Classification: LCC MT723 .A495 2018 | DDC 006.5--dc23
LC record available at https://lccn.loc.gov/2018022623

www.halleonardbooks.com

Contents

Acknowledgments

A series like this is never the work of one person, but rather a collection of the experiences obtained over the years from too many people to acknowledge here. Yet some deserve a special mention.

Dan Earley, my editor at Music Sales, who was the first person to say, "You know what would be cool? A series of books on recording, like those Time Life libraries." Well Dan, better late than never, right?

Sir George Martin, who was kind enough to write the foreword to my 1977 book, *Home Recording for Musicians*. He asked for samples of my writing, and I thought that would be the end of it. Instead, he wrote an eloquent foreword that set a wonderful tone for the book. He truly was the consummate gentleman everyone says he was.

The team at Hal Leonard—especially John Cerullo, who green-lighted this series and brought in Frank D. Cook to serve as the editor for these books.

My father, who taught me that it didn't matter if I was a dreadful writer as long as I could edit my words into something readable—and who also showed me what it meant to love music.

My mother, who with my father was unfailingly supportive when I wanted to do things like drop out of college, join a rock band, go on tour, and never look back!

My brother, who understood music on a very deep level and died too young.

And of course, the many *(many)* engineers and producers who let me look over their shoulders and absorb knowledge like a sponge over the past five decades. My hope is that this series will help pass their collective wisdom on to another generation.

Introduction

About this Book

Welcome to the book series Musician's Guide to Home Recording. This series of short publications was written to address the needs of musicians and recording enthusiasts who are interested in creating self-produced songs or doing audio production work for others.

Rather than trying to cover all aspects of recording in a single sprawling volume, each title in the series concisely and accessibly addresses a particular subject. You can select individual titles to hone in on certain skills or proceed through the entire series; this kind of approach lets you develop a comprehensive knowledge at your own pace.

This book, *How to Choose and Use Audio Interfaces*, covers the ins and outs (pun intended!) of audio interfaces, which are essential elements of today's computer-based recording.

The Purpose of an Audio Interface

Computers can be pretty smart, but they understand only digital data. Some digital instruments, like digital synthesizers, can feed data directly into a computer. However instruments like guitar, voice, saxaphone, piano, older analog synthesizers, acoustic drums, and anything else that generates an analog signal cannot feed a computer directly. That's where an audio interface comes into play.

An audio interface converts analog signals into data the computer can understand using an analog-to-digital (A/D) converter (also called an ADC), and converts data from the computer back into analog sounds we can hear using a digital-to-analog (D/A) converter (also called a DAC).

Most Windows and Mac computers have built-in audio capabilities used to produce sound effects in games, play the audio in streamed content, and so on. However, onboard audio typically uses consumer-grade devices. While these may be fine for casual applications, or while traveling with a laptop, you'll need an audio interface designed for professional audio applications to obtain the best sound quality and performance. Fortunately, a range of models exist to fit almost any budget—from something tiny that connects to your smartphone to industrial-strength models that can record a symphony orchestra.

Choosing an audio interface should be easy: decide what you need to do, and you will almost certainly find an interface that meets your needs (and hopefully, your budget). But how do you decide exactly what features you *do* need? The proliferation of audio interfaces has led to an overload of options. Which types of inputs do you need? Analog or digital? S/PDIF or AES/EBU? What about digitally controlled mic preamps?

While choices are always welcome, it can be difficult to separate what manufacturers want to sell you from what you really need.

This book will help you understand how audio interfaces work, recognize the importance of various features, and determine how to choose the model that will work best for you. Each chapter ends with a "Key Takeaways" section that summarizes the most important points in the chapter. And for those who want to dig deeper, I've included "Tech Talk" sidebars with additional information.

Rather than recommend specific models (new products are always being introduced), this book explains the concepts involved in audio interfaces so that you can become educated enough to make an informed choice to select the right interface for your needs—and then take full advantage of what it can do.

Tips and References

This book includes various tips, definitions, cross-references, and other supplemental nuggets throughout its pages. These are denoted with the following icons and formatting.

Tips and side notes provide helpful hints and suggestions, background information, or additional details on a concept or topic.

Definitions provide explanations of technical terms, industry jargon, or abbreviations.

Cross-References alert you to another section, book, or online resource that provides additional information on the current topic.

Chapter 1

Jacks and Connectors

One of the most important aspects of choosing an audio interface is identifying your recording needs; from there, you can decide how many inputs and outputs you need. For example, if you're a guitar-playing solo artist, you may need an interface with only two inputs—one for your voice and one for the guitar. But if you're miking a drum set, you'll want several microphone inputs. And if you're recording a live band, you'll need *a lot* of inputs. You also need to decide whether your inputs will come mostly from microphones, direct connections to devices like keyboards, or a combination of the two.

Balanced and Unbalanced Lines

Several types of audio connectors are used with interfaces. We'll get into the connector types in a moment. But first, you'll need to understand the difference between balanced and unbalanced connections, and the cable lines that run between them.

Unbalanced Connections

The typical musical instrument cable is an example of an unbalanced line with two conductors. One wire is the *ground* line (also called common, earth, shield, or DC return). The other is the *hot* line, which carries the signal. These typically use a two-wire 1/4-inch phone connector with a tip and sleeve (Fig. 1.1) or an RCA phono plug, as used in hi-fi gear (Fig. 1.2).

Figure 1.1 This 1/4-inch phone plug has two conductors for unbalanced systems. The tip is the hot connection, and the rest of the shaft (the sleeve) is the ground connection.

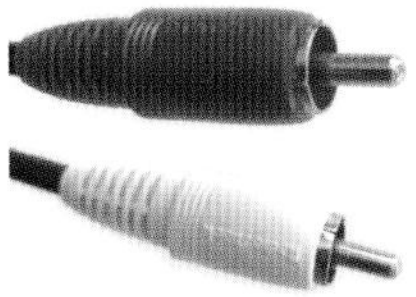

Figure 1.2 Phono plugs, also called RCA plugs, are unbalanced and used primarily in consumer gear.

Almost all guitar processors and amps use unbalanced inputs to be compatible with conventional guitars.

Balanced Connections

A balanced system adds another signal line (called the *cold* line) to the hot signal-line-plus-ground combination of an unbalanced system (Fig. 1.3). Although it's easy to geek out on the technical details, all you really need to know is that a balanced line helps reject hum and noise that a cable might pick up (such as when placed close to a transformer or AC power line).

Figure 1.3 This balanced phone plug has three conductors—the Tip (hot or +), the Ring (cold, –, or neutral), and the Sleeve (ground). The first letter of these terms is what gives balanced phone plugs and jacks their common abbreviation: TRS.

Tech Talk: How Balanced Lines Work

With a balanced cable, as an audio signal's voltage increases along one line, its mirror image on the other line decreases. This type of signal has to feed a special type of input, found in transformers and certain types of amplifier designs, that responds to the difference between these two out-of-phase signals. This *differential input* rejects signals that spill over into both signal lines equally, such as hum and noise. Because the differential input responds only to the differences between the two signal lines, when the same signal is present on both lines, the differential input ignores it.

Is Balanced Better?

Balanced lines are preferred with long cable runs, especially ones carrying low signal levels that are subject to noise or interference. In a small studio or onstage setup with relatively short cable runs and high signal levels, unbalanced connections are sufficient—but be sure to use balanced connections for your microphones, because almost all professional mics are designed for balanced-line operation.

Signal Level Categories

Audio interfaces need to handle three broad categories of signal levels.

- *Mic level* signals have the lowest level of the three, and therefore require amplification to bring them up to line level.
- *Line level* is the higher-level signal generated by electronic instruments, mixers, and other devices that don't require additional amplification within the interface itself.

- *Instrument-level* signals, associated with electric guitar, electric bass, and other instruments with passive pickups, have a signal strength somewhere between mic and line levels.

A guitar pickup is a transducer that converts the mechanical vibrations of the strings into an electrical signal. Passive pickups provide no electrical amplification for their signals, whereas active pickups include a built-in preamplifier, generally powered by a battery.

Tech Talk: Understanding the decibel (dB)

The decibel is a unit of measurement for audio levels (like an inch or meter is a unit of measurement for length). A 1 dB change is approximately the smallest audio level difference a human can hear. A dB specification can also have a – or + sign. For example, a signal with a level of –20 dB would sound softer than one with a level of –10 dB, but both would be softer than one with a level of +2 dB.

With line-level analog signals, two common nominal levels are used: +4 dB (commonly used in pro setups) and –10 dB (used more for home recording). For example, most guitar preamps, synthesizers, and other music gear use –10 dB levels. Professional, large-format mixing consoles use +4 dB levels.

In theory, connecting +4 and –10 gear together can create a gain mismatch. In practice, today's recording gear is relatively tolerant of different levels. Feeding a –10 output into a +4 input will likely not cause a significant noise increase. Conversely, simply turning down an input control can trim a +4 signal down to –10 levels. Some interfaces have a software or hardware switch that optimizes the interface for line-level signals at either +4 or –10 dB (Fig. 1.4 and 1.5, respectively).

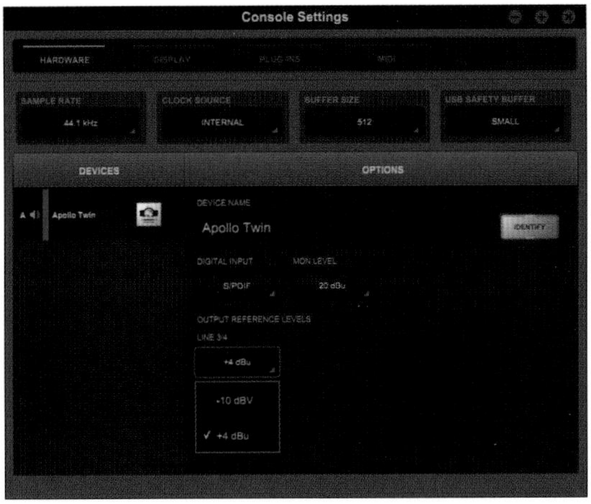

Figure 1.4 The software control panel for Universal Audio's Apollo Twin audio interface has a software switch for selecting an output reference level of either –10 or +4 dB.

Figure 1.5 TASCAM's US-20x20 interface has a rear-panel switch to select between nominal –10 dB and +4 dB levels.

Connector Types

Most audio interfaces have several types of connectors to allow connecting a variety of gear. Following are the most common analog connection types.

XLR Connectors

XLR connectors are used on professional–level microphones and some gear with +4 signal levels. XLR connectors have balanced connections with three pins (Fig. 1.6). The jacks usually indicate the associated pin numbers (1, 2, and 3) because there is a standardized way to wire these: pin 1 is ground, pin 2 is the hot or + connection, and pin 3 is the cold or – connection.

Figure 1.6 The XLR plug on the left connects into one of the XLR microphone inputs on the right.

Condenser mics need a source of power to operate, called *phantom power*. Most audio interface XLR jacks can switch on phantom power, which sends power from the XLR jack down the microphone cable to the condenser mic. You may be able to switch on phantom power for individual mic inputs, groups of mic inputs, or all mic inputs. Phantom power voltage is typically +48 Volts, but this isn't super-critical; most condenser mics work just as well at lower voltages.

Connecting a dynamic mic that uses a balanced line into an XLR connector that generates phantom power will *usually* not cause problems. However, *devices with XLR outputs that aren't microphones should not be subjected to phantom power*. Although their outputs may be protected against accidentally applying phantom power, this isn't guaranteed.

1/4-Inch Phone Connectors

The jacks used for 1/4-inch phone connectors look like the jacks on guitar amps (which usually use unbalanced connections), but they may not all be electrically identical because they can be balanced or unbalanced. These jacks accept outputs from synthesizers, drum machines, and similar devices that have line-level outputs.

The jacks on 1/4-inch balanced TRS phone cables appear similar to standard 1/4-inch jacks for a guitar cable; however, balanced connections use stereo jacks that are wired for balanced operation.

See the discussion earlier in this chapter on balanced and unbalanced lines for illustrations of the types of plugs used with these jacks.

Because it's not obvious when looking at the outside of phone jacks whether they're balanced or unbalanced, companies will often include an indication that they're balanced, and may even show the wiring (Fig. 1.7).

Figure 1.7 The line outputs on this PreSonus interface use balanced TRS connections. Most interfaces designed for home recording use 1/4-inch TRS output jacks for balanced lines because there isn't enough room to dedicate XLR jacks on both the inputs and outputs.

Note that balanced 1/4-inch connections are compatible with 1/4-inch unbalanced lines because if you use an unbalanced plug, a balanced output simply functions as an unbalanced output. However, the level will be half as much compared to a balanced version of the same signal. Also note that adapters are available to convert XLR outputs to balanced 1/4-inch outputs, and vice-versa.

Combination Jacks

Combination jacks (or combi jacks) are common on compact, all-purpose audio interfaces. These space-saving jacks combine an XLR connector with a 1/4-inch TRS connector, located inside the XLR section (Fig. 1.8). This universal solution lets you plug in an XLR balanced cable, a 1/4-inch balanced cable, or a 1/4-inch unbalanced cable.

Figure 1.8 A combi jack takes up about as much space as a standard XLR jack, but you can insert a 1/4-inch phone plug into the center.

1/8-Inch Stereo Mini Connectors

While waning in popularity, you'll find these smaller connectors in some smartphones, extension speakers, MP3 players, and the like. Sometimes they're also used as output jacks designed to feed headphones (Fig. 1.9).

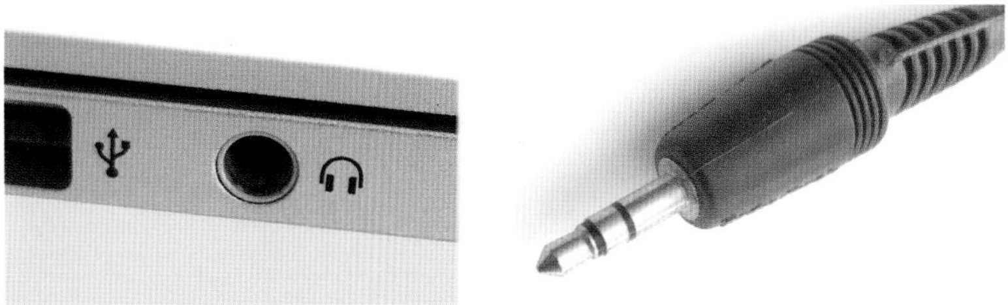

Figure 1.9 1/8-inch headphone output jack (left) and 1/8-inch stereo plug (right)

> If you need to use a stereo 1/8-inch input or output and no matching cable is available, you can use a 1/4-inch to 1/8-inch adapter with a standard 1/4-inch stereo cable.

Instrument Inputs

Guitars, basses, and other instruments with passive pickups (i.e., pickups that don't include an onboard preamplifier) have special requirements because they're not fully compatible with either XLR mic inputs or 1/4-inch line level inputs. So, most audio interfaces include a special 1/4-inch jack called an instrument, guitar, direct, or high-impedance (Hi-Z) input. This type of input helps preserve a pickup's high-frequency response and output level (Fig. 1.10).

Figure 1.10 These two mic inputs can also be instrument inputs because they use combi jacks and have switches to convert the inputs for use with high-impedance sources.

Tech Talk: About Impedance

When routing a signal to an audio interface input from a passive guitar pickup, some of the signal gets lost along the way—sort of like having a leak in a water pipe. Whether this leak is a pinhole or something bigger depends on the input's *impedance.* You can think of impedance as adding the electrical equivalent of friction between the audio signal and ground—the lower the impedance, the more signal loss occurs through the leak to ground. With stock guitar pickups, lower input impedances load down the guitar, which reduces the output and produces a "duller" sound.

How Guitar Cables Interact with Instrument Inputs

Cables can affect the sound of guitar, bass, and other instruments that use passive pickups, because cables have *capacitance*—the same electrical characteristic as a guitar's tone control capacitor.

Essentially, a cable applies a second, subtle, "always-on" tone control that reduces the treble. The cable capacitance has more of an impact if the impedance is high: with higher input impedances, it's more likely that cable capacitance will reduce the high frequencies somewhat.

For the brightest possible guitar sound, you'll want to use a cable with minimal cable capacitance. Note, however, that some guitarists have been known to play onstage using long cables, resulting in enough cable capacitance to reduce highs. This can create a "creamier" sound because it prevents subsequent distortion stages from distorting the high frequencies, which can create a brittle sound.

Cable capacitance is specified in *picofarads* (pF) *per foot*, where higher values represent greater capacitance.

One studio trick to emulate the effect of a long cable is to use a coil cord cable (Fig. 1.11), which generally has more capacitance than a non-coiled type. If you go this route, however, make sure to select a quality coil cord; cheaply made ones can generate noise when a performer moves around.

Figure 1.11 This coil cord from Vox is well made and comes in five different colors.

Audio Interface Inputs that Emulate Amps

An audio interface's instrument input is not necessarily electrically identical to a typical amp input; many guitarists feel an instrument input can't provide the same subtle interaction that happens between passive pickups and a physical guitar amp.

To compensate, some audio interfaces (such as the Avid Eleven Rack and Universal Audio interfaces with so-called Unison Technology) include circuitry designed to emulate an amp input. To get a similar effect, Radial Engineering makes direct boxes with a "Drag" control that emulates the loading of cables and lower impedances; these can be inserted between a guitar and an interface input. Likewise, MOTU's Zbox is an outboard device that inserts between a guitar and interface to give more of an amp "feel" (Fig. 1.12).

Figure 1.12 MOTU's Zbox is designed to give more of an amp feel when inserted between a guitar and a high-impedance audio interface input.

In most cases, using a quality, low-noise cable between your guitar and audio interface will do what you need. However if you're a guitarist who is used to a cable affecting your sound, or if you depend on the interaction between your guitar and an amp, try some of the solutions above as you seek the perfect tone. Also remember that plugging into a mic or line input will degrade a passive pickup's sound.

Breakout Cables

With audio interfaces, panel space is at a premium. It's not always possible to fit all the inputs and outputs that an interface can handle on the front or back panel, especially with compact, mobile interfaces. The solution is a breakout cable—a multiconductor cable that connects to a small, multipin connector on the audio interface (Fig. 1.13).

Figure 1.13 This breakout cable plugs into a 9-pin connector on the back of an audio interface. It provides two 5-pin MIDI jacks and two RCA jacks that could not fit on the rear panel. Breakout cables can involve analog or digital connections, and typically use either 9-pin or 25-pin multipin connectors.

For examples of breakout cable connections, refer to Figures 2.6 and 5.2.

Key Takeaways

- Analyzing your recording needs will help you decide how many analog inputs you need in an interface and the type of connectors that best match your gear (although you can always use adapters if needed).
- Combi jacks are useful because they accommodate 1/4-inch balanced or unbalanced plugs as well as XLR plugs.
- Instruments with passive pickups require high-impedance instrument inputs to avoid reducing level and treble.
- You don't need to fret too much about +4 and –10 dB levels, or balanced and unbalanced lines, unless you have a big studio with long cable runs. You'll likely use –10 dB.

Chapter 2

Audio Interface Types and Computer Connections

Audio interfaces come in various forms and connect to computers in various ways. This chapter covers the different interface packaging, computer connections, and driver software that allows a computer to talk to the interface.

External Box

The simplest type of interface to install is a box that's external to your computer with several analog inputs and outputs, digital inputs and outputs, and a jack to connect a cable to a *computer port*. This is a common type of professional audio interface (Fig. 2.1).

Computer port: A computer connection that exposes the computer's internal circuitry to the outside world. Examples include USB, FireWire, and Thunderbolt, discussed later in this chapter.

For detailed information on digital connections, see Chapter 5 in this book.

Figure 2.1 Three compact USB interfaces that connect to a computer via a single USB cable: (left to right) ESI U22 XT, TASCAM US-2x2, Focusrite 2i2.

The interface converts your analog audio signal to digital audio and then sends it to the computer via a digital cable connection (like USB, Thunderbolt, or FireWire—we'll cover these options in some detail shortly). Unlike a standard audio connection, this single connection cable is *bi-directional*—it can send your input signals to the computer and also receive output data from your computer (e.g., the mixed output of a recording program).

The main advantages of an external box interface are:

- You don't need to open up the computer to install anything.
- Sensitive audio circuitry lives outside of the electrically "dirty" computer environment.
- You can use the interface with other computers if needed.

However, these devices also have some constraints. For best results you need to connect the interface to a port on the computer itself, not a port in an external *hub* (an expansion device with multiple port connectors) (Fig. 2.2). Some hubs are passive while others are powered, but neither type is recommended for audio interface connections.

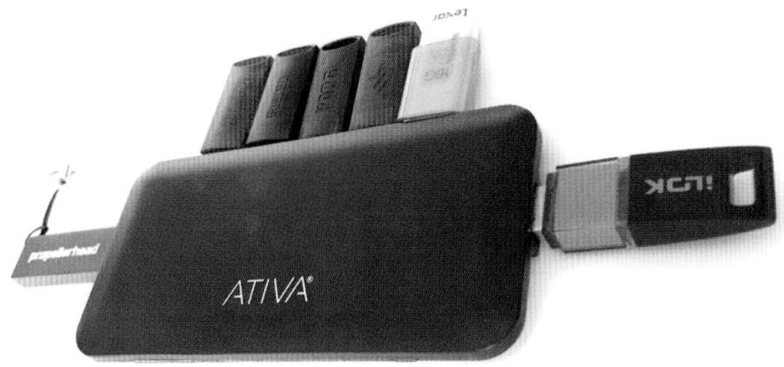

Figure 2.2 This USB hub provides multiple ports from a single computer USB port.

External Mixer with Audio Interface

Some audio interfaces look like the conventional mixing consoles used for live performance, but can connect to computers. (See Fig. 2.3.) This is an ideal interface solution when you want to record a live gig.

Typical mixers/audio interfaces can send the mixed signal out for recording but can also send individual channels to recording software for editing, mixing, and polishing a live performance in the studio. Some can even record the raw tracks directly to a hard drive or qualified USB memory stick, so you don't need to take a computer to the gig.

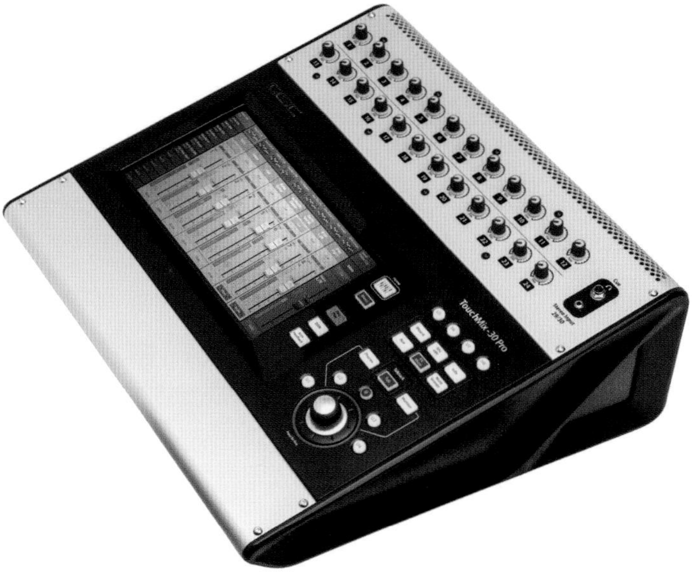

Figure 2.3 QSC's TouchMix-30 doubles as a live performance mixer and a studio interface with inputs and outputs for each channel, and offers the option to use an iPad remote for control.

This type of interface can make sense for solo artists too, because *all* their instruments (guitars, basses, synthesizers, drum machines, microphones, etc.) can be plugged in and have their levels optimized. Each instrument can be selected instantly inside recording software instead of requiring re-patching with a limited number of inputs.

External Gear with Built-In Audio Interface

Many guitar processors (including multieffects from Line 6, DigiTech, Roland, Zoom, and others) as well as some digital keyboard "workstations" and even guitar amplifiers include audio interfaces (Fig. 2.4).

Figure 2.4 The Line 6 FireHawk 1500 amp can also function as a basic audio interface—note the USB port.

Some of these interfaces are surprisingly capable and may be all you need for computer-based recording. Before looking to buy a dedicated audio interface, check whether gear you already have can do the job.

DJ Controller with Built-In Audio Interface

DJ controllers designed to work with computers often include audio interfaces (Fig. 2.5). These may not be as comprehensive as a dedicated audio interface designed for multitrack recording, but nonetheless they often include inputs, outputs, and typically a mic input.

Figure 2.5 Native Instruments' Traktor Kontrol S2 is a basic DJ controller that provides main and booth outputs, as well as separate cue channels and a mic input.

Internal PCIe Card

Back when most computers had internal slots for adding functionality from a graphics card, Ethernet card, modem card, etc., an internal audio card was a common type of audio interface. PCIe cards are still used in high-performance applications because they connect directly into the computer's data bus instead of using a data port. This direct connection provides more efficient operation than that of an external box (although practically speaking, the differences are relatively minor with modern, high-speed connection protocols).

The card's *backplate*—the side that faces out from the computer—will typically have several jacks for analog audio, or digital input and outputs (described below) to accommodate digital gear. Because the backplate doesn't have a lot of space, some cards instead use a multipin connector (Fig. 2.6) for a breakout cable with various connectors. Alternatively, the setup may include a multipin cable that connects to a separate box with various connectors.

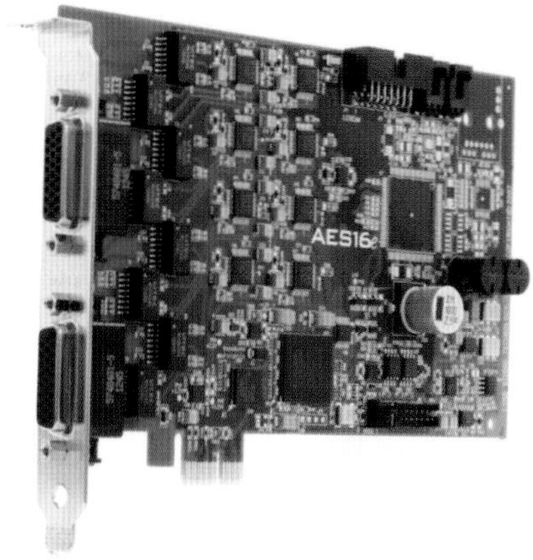

Figure 2.6 This PCIe interface from Lynx Studio Technology supports up to 16 channels of AES/EBU-compatible digital audio.

PCIe cards that include analog inputs and outputs directly on the backplate present a more challenging design because computers tend to be *electrically dirty*. It takes careful circuit board layout and shielding to make sure that spurious computer signals don't affect the audio quality (Fig. 2.7).

> *Electrical dirt*: Computer operations generate considerable electrical noise, some of which is in the audio range. These signals can cause interference that negatively affects audio quality.

Figure 2.7 ESI's MAYA44 eX includes four analog inputs and outputs, as well as an optical port for S/PDIF digital connections.

Although you may experience higher performance with PCIe cards, there are disadvantages too—you need to open up the computer and install the card, and if the card handles analog audio, it can pick up noise from the computer.

Note that modern Apple computers no longer include slots for PCIe cards. Although it is often possible to use adapters that accept cards, an external interface may be much simpler and can provide excellent performance.

Digital Connections for External Boxes

Three main data connections are available to hook up an external box to your computer: USB, FireWire, and Thunderbolt. All of these options provide a bi-directional data bus and can provide power to a connected peripheral.

The amount of power provided by the data bus is often enough so that the audio interface doesn't require an external power supply; these types of audio interfaces are called *bus-powered.* However, many bus-powered devices also allow for external power in case the bus can't provide enough power for the interface or the bus is powering other devices as well. Not all computers have ports for the three main connection options, so make *sure* you know what your computer uses as you consider what type of interface to buy.

An important characteristic of all digital connections is *speed*—how fast they can transfer data. In practical terms, higher speeds have two main benefits:

- They allow handling a greater number of audio channels simultaneously (for example, in a live recording you may need to record multiple signals at once).
- They have less *latency*. Latency is an important topic we'll discuss in depth, but the basic idea is that when you monitor a signal through the computer, there can be a noticeable delay. For example, if you're singing a vocal and monitoring the computer's output in your headphones, with higher latencies you'll hear a delay, as if you were singing through an echo unit. With high-speed interfaces, this delay will likely not be objectionable or even noticeable.

USB (Universal Serial Bus)

This is the most common option for external boxes, because it gives good performance and is generally very reliable. USB has evolved over the years to become faster and more efficient. Early USB 1.1 interfaces supported a very limited number of audio channels. USB 2.0, the next generation, used the same connectors for physical compatibility but offered a significant speed increase and allowed interfaces to expand their channel count. USB 3.0 is somewhat of a special case because it added more lines for greater speed, thus changing the connector; however it remained compatible with USB 2.0 connectors if you were willing to forego the speed increase. USB 3.1 and 3.2 upped the speed even more, and changed the connector again to the more universal USB-C type.

Using suitable adapters, older USB devices can connect to USB-C ports.

Until USB-C came along, most connectors that plugged into the computer were the same—the USB Type A. However, a cable with a Type A connector at one end could have any one of a number of connectors at the other end for connecting to peripherals like audio interfaces (Fig. 2.8).

Figure 2.8 Left to right: USB Type A connector, USB Type B connector, USB Mini B connector, USB Micro B connector, and USB-C connector.

The USB Type A connector is rarely used with peripherals, because it almost always plugs into a host device. The USB Type B connector, which is designed for larger USB peripherals, is the most common connection for non-mobile audio interfaces. The smaller USB Mini B connector is an older choice for mobile-friendly audio interfaces; it's no longer being designed into new devices, but is still in relatively wide use (especially with cameras).

The USB Micro B connector is a more popular choice for mobile musical peripherals, and is commonly found on smart phones as well. The USB-C connector is the main connector used for USB 3.1 and USB 3.2 ports. This connector is extremely thin and is also reversible—you don't have to orient the plug in a particular way to plug it in.

USB 3.0 interfaces offer improved performance over USB 2.0 devices and will remain common for some time. The end of the cable that connects to the host looks like a standard Type A connector, however it uses blue plastic to differentiate it from cables that are not compatible with USB 3.0. Because USB 3.0 needs more connection points, the end that goes to peripherals has an extra section added on top. This is why USB 3.0 plugs (Fig. 2.9) don't look like the standard plugs for USB Type B connectors.

Figure 2.9 USB 3.0 plugs are usually identified as either USB 3.0 or "SuperSpeed" USB. Note how the plug resembles the older Type B, but has extra connections stacked on the top.

So what does this all mean? Here's a summary.

- USB 1.1 still works, but it has limited speed and no newer products use it.
- For most applications, USB 2.0 interfaces are all that's needed and are almost always compatible with USB 3.0 as well as USB 3.1 ports, so you can continue using your interface if you update your computer. However the interface will still operate at USB 2.0 speeds even if connected to a high-speed port, and you will need an adapter for use with USB-C ports.
- USB 3.0 interfaces are fast and efficient. Although some of the earliest models had compatibility issues with early USB 3.0 chipsets, compatibility is no longer a factor except with older computers.
- As of this writing, no USB 3.1 interfaces exist. It remains to be seen whether USB or Thunderbolt interfaces (described later) become more popular in the years ahead.

> Thunderbolt is a faster protocol, but it is more expensive. It's likely that Thunderbolt will be the choice for higher-priced interfaces, while USB 3.1 will be the next step in USB interfaces for general home recording.

FireWire (aka IEEE-1394)

This is a common, but fading, connection protocol. Although intended originally for use with peripherals like camcorders, FireWire can work well for transferring digital audio between a computer and an external interface. If your Windows computer lacks FireWire but has PCIe slots, you can add a FireWire PCIe card that provides multiple FireWire ports. (Note, however, that several manufacturers recommend against using combination FireWire/USB cards.)

If your computer doesn't have slots but has a Thunderbolt connection (described below), FireWire audio interfaces will work using an adapter, although their performance won't match Thunderbolt's. Because

FireWire is not as popular as it once was, it's possible to pick up high quality, previously owned FireWire audio interfaces at low cost.

As with USB, FireWire has evolved over the years. The original FireWire 400 interface has two types of connectors: four-pin and six-pin. Aside from size, the main difference is that the six-pin connector can provide power as well as carry data, whereas the four-pin connector carries only data. The subsequent FireWire 800 specification provides twice the speed of FireWire 400 and uses a connector with nine pins (Fig. 2.10).

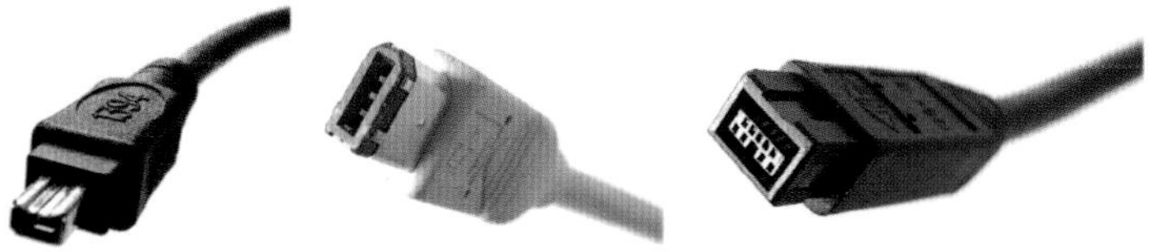

Figure 2.10 Left to right: Four-pin and six-pin FireWire 400 connectors, and FireWire 800 connector. Some FireWire 400 systems use the FireWire 800 connector.

Because the four-pin connectors do not provide power from the bus, you must use an external AC adapter for any peripheral that uses this option. Although some six-pin to four-pin adapters exist, these carry only signal—they don't magically provide power to four-pin devices.

Although FireWire devices are supposedly *hot swappable*—meaning you can plug and unplug them from a computer while the power is turned on—you should exercise caution here. There have been isolated reports of motherboards failing from having FireWire peripherals plugged in while both the audio interface and the computer were powered on. It's always safest to make connections with the computer and peripheral turned off.

Thunderbolt

Thunderbolt ports are most commonly associated with Apple computers, although you'll find them more and more on Windows machines. Thunderbolt 3 is the latest version and offers several advantages over previous generations (as well as other interface protocols), including higher speed. Whether this matters for your audio interface depends on how important low latency is to you.

Refer to Chapter 3 of this book for details on latency—what causes it and how to minimize it.

Adapters are also available for devices using FireWire, PCIe cards, earlier versions of USB, and other protocols so that they can be compatible with Thunderbolt ports. Thunderbolt 3 ports are backwards compatible with almost all older Thunderbolt peripherals. Additionally, Thunderbolt 3 peripherals may be compatible with Thunderbolt 1 or 2 ports.

Before investing in a Thunderbolt interface, *make sure your computer is compatible.* The interface manufacturer's web site will likely provide information on compatibility, what has been tested, what hasn't been tested, and the like. For example, an interface may be supported starting with a particular processor generation or Thunderbolt 3 controller chip, but it may not be tested with previous generations. Just because something isn't officially supported doesn't mean it won't work, but you should be aware that you're taking a chance.

Thunderbolt 3 uses the same USB-C connector as USB 3.1, and Thunderbolt ports are compatible with USB 3.1 peripherals. (However, the reverse isn't true—you can't hook up a Thunderbolt peripheral to a USB 3.1 port.) The first two generations of Thunderbolt used a different connector type and were not natively compatible with USB (Fig. 2.11).

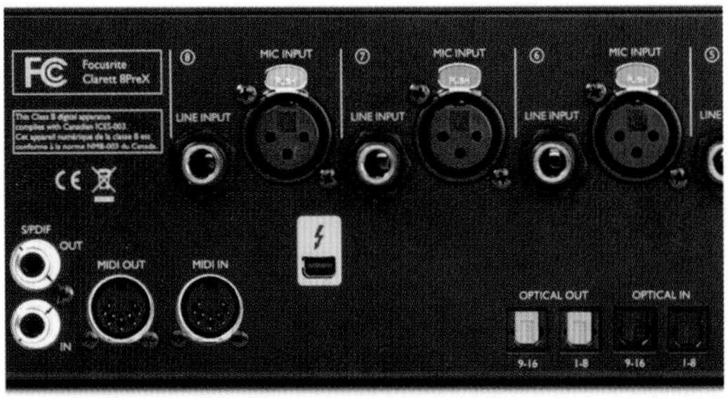

Figure 2.11 Thunderbolt 1 and 2 used a different connector compared to Thunderbolt 3's USB-C connector. Focusrite's Clarett 8PreX groups the Thunderbolt port with other digital inputs and outputs (S/PDIF, MIDI, and optical).

Note that Thunderbolt 3, as well as USB 3.1, can deliver up to 100 Watts of power (the exact amount depends on the port's power profile). It's also possible to *daisy-chain* multiple Thunderbolt peripherals to one another and connect them all to a single port on the computer (Fig. 2.12).

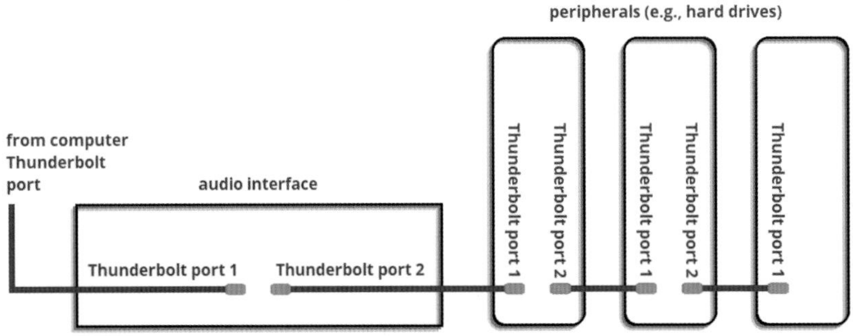

Figure 2.12 Daisy-chaining can connect up to six Thunderbolt peripherals together as one long chain.

Interface Cables

Back when interfaces were relatively slow, just about any cable with the right connectors would work between your interface and computer—but times have changed.

With USB 3.0 and higher, you need cables that are designed specifically to handle the protocol's higher speeds. And if you're planning on using USB-C–compatible cables to carry significant amounts of power, you need to make sure that the cable you choose can handle the power—the skinny little USB-C cord that you use with your smartphone probably isn't up to the task.

Thunderbolt can use two different types of cables. Passive cables use ordinary wire, and the transfer speed slows down with lengths beyond half a meter. Active cables include electronics built into the cable to regulate data transfer. These can maintain higher speeds over longer distances—but as you've probably guessed, they're more expensive. Whether the extra performance is worthwhile depends on your requirements, your budget, the distance the signal needs to travel, and the type of Thunderbolt port you're using.

Interface Drivers

Drivers are software programs that facilitate communication between your computer and audio interface. They are the data's gatekeepers; how efficiently they do their task affects latency (delays through the computer) and overall performance. Mac and Windows computers each have their own driver formats.

Driver Control Panels

Drivers typically have control panels for setting various parameters that we discuss later, such as sample rate, bit resolution, and sample buffers. You can generally open the control panel directly from your Windows or Mac operating system, but there will usually be a way to access it from within a host program as well.

The host will communicate with the interface based on the parameters that the control panel specifies (Fig. 2.13).

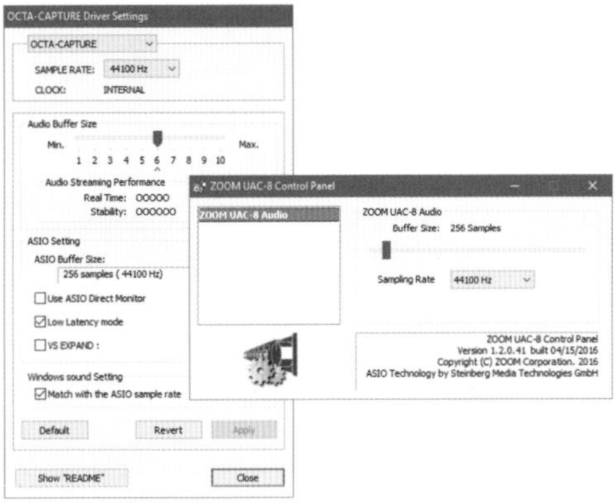

Figure 2.13 Typical control panels (for the Roland Octa-Capture and Zoom UAC-8 interfaces).

Class-Compliant Interfaces

Some interfaces are labeled as *class-compliant.* This means that they require no special drivers to work with computers and are truly plug-and-play. However, custom drivers will often give better performance or improved features, especially with Windows computers.

Certain interfaces include a switch to select class-compliant operation (Fig. 2.14) if you're using the interface with a computer that doesn't have drivers installed.

Figure 2.14 Turning on class-compliant mode allows using an interface with any computer, even if no drivers are installed.

Windows

The Windows operating system includes various drivers developed by Microsoft for audio applications like game sound and music playback. These include the MME driver (it's ancient and slow—avoid it), the WDM/KS driver (this never really caught on), and the modern, low-latency WASAPI driver optimized for internal sound chips. The latter is an excellent choice for laptops, assuming your software supports the low-latency modes.

Although most interfaces and software programs can work with Microsoft's drivers, the ASIO driver created by Steinberg Media Technologies remains the preferred, and most commonly used driver for Windows pro audio and recording. When you select a driver in your recording software, you'll see any drivers associated with the interfaces you've used or are using (Fig. 2.15). A custom ASIO driver written by a manufacturer for their specific audio interface will be the best choice compared to using a native Windows driver.

ASIO stands for Audio Stream Input/Output.

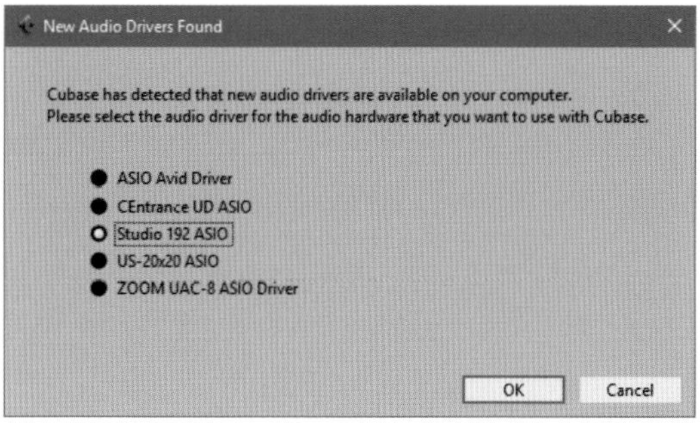

Figure 2.15 If you have several interfaces, or have used other interfaces in the past, their drivers will be options when selecting an audio device for your software. It's good practice to uninstall drivers you no longer use.

In some cases, a manufacturer will not have taken the time or effort to create a custom driver. In that event you have to use either the Windows drivers or an alternative, free driver called ASIO4ALL. The goal of the ASIO4ALL driver is to convince ASIO audio interfaces that standard Windows drivers are ASIO. It's a clever kludge that can reduce latency, but while it may work well in some systems, it doesn't work well in all systems. In some cases, ASIO4ALL can interfere with other installed drivers. Most pros do not rely on it, although it can be handy when you need lower-latency, non-critical performance with laptops.

Apple Mac

The Mac standardized on *Core Audio* drivers with OS X, thus avoiding the kind of confusion that can occur with Windows. Although some interfaces require custom applications to accommodate particular unique features, modern Mac interfaces will be compatible with OS X out of the box.

Driver Installation

Driver installation is more of an issue with Windows computers; be sure to read and follow any installation instructions carefully. Although recent devices are mostly plug-and-play, some drivers require installation with the interface hooked up to the computer, while others require connecting the interface to the computer after installation (Fig. 2.16).

Figure 2.16 Follow installation instructions carefully. Line 6 displays an installation screen to advise you that you need to connect the interface *after* installing the drivers—a common practice for USB peripherals.

Other Software Used with Interfaces

Other small programs may also accompany the driver, like a virtual mixer for routing interface inputs and outputs within the computer (Fig. 2.17). These may need to be installed individually, or they may be installed as part of a suite along with the driver itself.

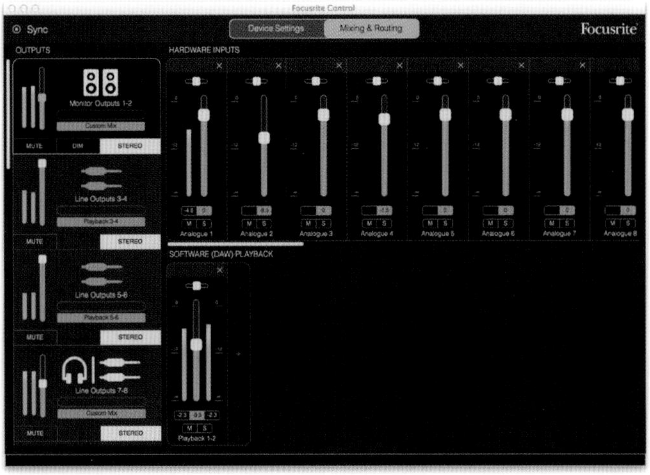

Figure 2.17 The Focusrite Control mixer/settings app for the Clarett8 PreX has a streamlined look. It allows you to adjust input and output levels, but you can also access the Device Settings by clicking on a different tab.

Driver Installation Tips

Driver installation is pretty painless. However, with older systems and some configurations, you may run into difficulties. Here are some tips.

- It's good practice to unplug all unneeded USB devices (in other words, everything except keyboard and mouse) prior to installing drivers for a USB interface.
- If during installations on Windows you're warned that the driver has not passed Windows logo testing, continue with installation anyway. This simply means the manufacturer decided not to pay for the tests Microsoft requires to certify operation.
- If the USB device has internal memory (e.g., for storing configurations), it may show up as an external computer drive. Before physically unplugging the USB device, eject it properly.
    - *Mac:* Drag the USB drive icon to the trash (which turns into an eject button) and wait several seconds before actually removing the USB connection.
    - *Windows:* Click on the Safely Remove Hardware button in the taskbar; then click on the USB device you want to eject (Fig. 2.18).

Figure 2.18 Click on the Windows Safely Remove Hardware taskbar button to prevent potential loss of data and other problems when disconnecting USB devices.

- Unless the computer is communicating with the USB device, no negative consequences will likely result from disconnecting the device without ejecting it first (other than seeing a scary warning on your computer screen). However, it's best to use the correct disconnection protocol.
- Audio interfaces should be connected directly to a USB port on the computer itself, not through a USB hub—especially an unpowered hub. Hubs are suitable for low-bandwidth devices like mice, QWERTY keyboards, printers, MIDI keyboards (that don't include an audio interface), and copy-protection dongles.
- The USB ports built into computers may not be suitable for high-performance audio; for example, they may introduce noise because they're shared with other peripherals. With Windows desktop computers, I've solved some stubborn USB problems by adding a PCIe card with multiple USB ports and dedicating its ports only to audio.
- If all else fails, check for updates to your motherboard and BIOS. Intel makes this particularly easy—search for Intel Driver & Support Assistant, download it, let the program analyze your computer, and it will alert you to any available updates.

Example: I had a computer whose USB 3.0 ports worked with some USB 3.0 devices but not others. After doing a firmware update, they worked with every USB 3.0 device I tried.

- With some drivers on Windows, if you plug a USB device into a port other than the one used during installation, Windows may not recognize the device. It then asks if you want to install the drivers. This can happen with older or less-than-rigorously-written drivers. Usually Windows will re-install the driver automatically, but if not, you can try re-installing the driver without the interface connected to the USB port. If Windows still doesn't recognize the interface, try re-installing the driver with the interface connected.

Updating Drivers and Related Applications

Drivers and any related applications (like software mixers that route interface inputs and outputs) are very important. Manufacturers update them occasionally to improve performance, or to ensure compatibility with operating system changes. It's a good idea to check a manufacturer's web site periodically for updates, even if you bought a product the day it hit the stores. Registering an interface may mean you'll receive email notifications for updates, but that's not guaranteed (and notifications may end up in your email program's spam folder, so check there periodically).

Sometimes the web site will be confusing, because it will list drivers and applications for multiple operating system revisions and different hardware (e.g., 32-bit as opposed to 64-bit computers). Make sure that whatever you download and install is compatible with your hardware and operating system.

Evaluating Drivers

You'll often see comments on Internet forums about certain interfaces that have "rock-solid" drivers (which usually translates to, "It works fine on my system") and others that have "flakey" drivers (which usually translates to, "It doesn't work so well on my system"). It's difficult to evaluate driver performance, yet there are definite performance differences among them. Furthermore as drivers are updated, today's under-performer might be tomorrow's star. Your only real option: Check out discussion forums, and look for threads about driver functionality on different platforms and with different interfaces. Although reports will vary, sometimes you'll find a theme that a particular company's interface drivers are better than average.

Uninstalling Windows Programs

If you no longer use an interface, there's no need to keep its driver installed. Uninstallation can generally be accomplished in one of three ways (Fig. 2.19):

- Using an uninstall option in the Start menu
- Choosing the uninstall command in the Device Manager (type Device Manager in the Task Bar search box to open it, right-click on what you want to uninstall, then choose uninstall)
- Uninstalling in the Add/Remove Programs section of the Control Panel

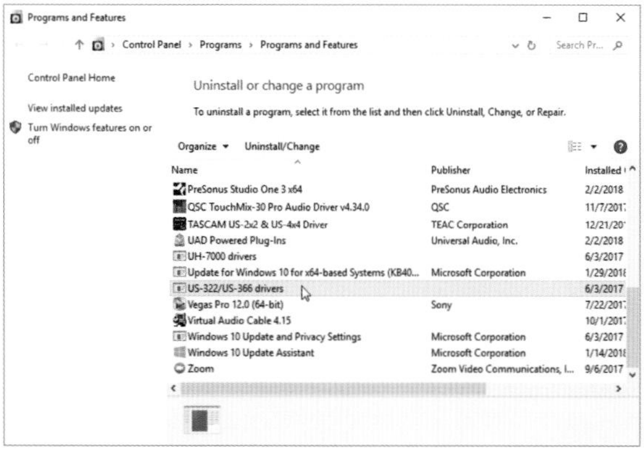

Figure 2.19 Different Windows programs uninstall in different ways. The drivers from an older TASCAM interface can be uninstalled via the Windows control panel.

Uninstalling Mac Programs

Apps in folders on the Mac may have an uninstaller. If so, double-click on the uninstaller and follow any onscreen instructions to uninstall the program. If no uninstaller exists, drag the app from the Applications folder to the Trash icon in the dock. Empty the trash to remove the program completely.

The Future

New computer protocols and standards are constantly being developed. Several generations of USB, FireWire, and Thunderbolt have been developed, with each one improving on the last.

In general, if your interface works today, it will give the same performance tomorrow. The one exception is drivers that become incompatible with an updated operating system. In that case, the interface may begin to work poorly, or not at all. On Windows, however, even though an operating system update may *seem* unable to recognize a driver, simply re-installing the driver can sometimes restore everything to working condition.

How to Choose Driver Settings

As mentioned previously, your audio interface will likely include its own control panel where you can adjust parameters that affect performance, compatibility, and other characteristics. This control panel will often be accessible from a Preferences or Options menu in your host software so that you can control the interface from within the program you're using.

The main characteristics you'll adjust for your interface include the following:

- *Sample rate*—This sets the rate for converting analog signals to digital representations and vice-versa.
- *Bit resolution*—This determines the accuracy with which you can record audio signal levels.
- *Latency*—This determines how much delay you'll hear when monitoring signals through your computer.

We'll investigate these settings further in the next two chapters.

Key Takeaways

- An external interface that connects to the computer via a cable is the most convenient solution.
- It never hurts to have more inputs than you think you'll need.
- USB 2.0 interfaces are fine and cost-effective for the majority of home recording situations; USB 3.0 interfaces give a bit of a performance boost.

- FireWire still works but is becoming less common. You can often pick up good deals on used FireWire interfaces.
- If you need the highest possible performance, choose a Thunderbolt interface—but make *sure* your computer has a compatible Thunderbolt port.
- On Windows, ASIO drivers written by the interface manufacturer give the best performance.
- Follow any instructions on driver installation to the letter, and check for updates regularly.

Chapter 3

Choosing Settings for Sample Rates and Bit Resolution

Audio interfaces work in tandem with your recording software, so they both have to use and support the same audio format characteristics. One key format parameter for a digital audio file is the sample rate it uses. Another is the file's bit resolution (or bit depth).

Sample Rates

You will set the sample rate for your recording either in your audio interface's control panel or via your recording software's preferences (Fig. 3.1).

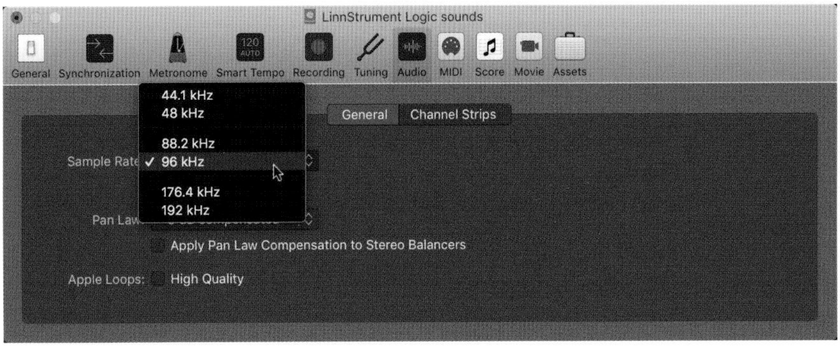

Figure 3.1 With Apple Logic, the File > Settings > Audio Preferences section is where you choose a project's sample rate. The process for setting the sample rate varies from one program to another, but it's usually associated with audio preferences.

About the Conversion Process

Sorry to get all nerdy on you, but understanding a bit about digital audio theory will help you make intelligent decisions when setting up your computer-based studio.

Analog-to-digital conversion translates real-world audio into computer data, so the more accurate the conversion, the better the audio quality. (Digital-to-analog conversion is equally important, but technically somewhat easier to do.) Conversion takes advantage of the fact that all acoustic sound, from a dog bark to a symphony orchestra, is simply a series of changes in air pressure. A dynamic microphone has a diaphragm that moves back and forth as it responds to air pressure. Because the diaphragm is positioned in a magnetic field, this movement generates a voltage in a coil. The changes in voltage correspond to the changes in air pressure.

Because a computer can't understand a changing voltage unless it's presented as a series of numbers, an *analog-to-digital converter* (ADC) measures the incoming analog voltage thousands of times a second and converts each measurement into computer-friendly digital data (binary numbers). The number of measurements the converter takes each second is the *sample rate* (Fig. 3.2), also called *sampling frequency*.

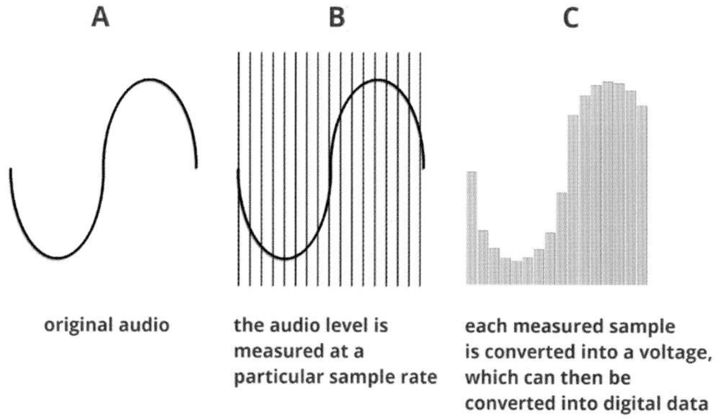

Figure 3.2 An analog signal (A) has a sample taken of its voltage at a certain rate (B) by an analog-to-digital converter, which produces a voltage (C) that's converted into digital data.

Common Sample Rates

CD-quality audio measures the incoming voltage 44,100 times per second (a 44.1 kHz sample rate) and can reproduce frequencies in the audio range (20 Hz to 20 kHz). However, it's not the only available sample rate. Here are some other common choices:

- **Lower than 32 kHz.** These sample rates are not sufficient to create quality audio. When you hear an electronic greeting card or toy, the sample rate will be well below 32 kHz.
- **32 kHz.** This is used in broadcasting and many satellite transmissions.
- **44.1 kHz.** Most home recording projects use this because it's the "sweet spot" between good fidelity and computer performance.
- **48 kHz.** Although not used much for purely audio projects, video projects often use 48 kHz for audio. Some people think 48 kHz sounds better than 44.1 kHz, but any difference is subtle at best.
- **88.2 kHz.** Many people claim this sounds better than 44.1 kHz, while others don't hear much of a difference. 88.2 kHz is much less prevalent than 96 kHz.
- **96 kHz.** This is the most popular "high-resolution" sample rate, and is common in DVDs and other high-end audio recording processes. Like 88.2 kHz, if you can hear a difference and your gear is up to

the task, there's no real reason not to use it, and there are situations (see below) where it can make an audible improvement.

- **176.4 kHz and 192 kHz.** Although some audiophiles believe audio at these ultra-high sample rates sounds better than audio at 44.1 kHz or even 96 kHz, these sample rates require considerable extra storage and stress computers more, so they're not used much for multitrack recording projects.

The "High-Resolution" Sample Rate Controversy

Fight club time! There's much debate about whether sample rates higher than 44.1 kHz yield a significant sonic improvement. Theoretically, higher sample rates aren't necessary. Still, some people insist that higher sample rates are better.

I've recorded projects at both 44.1 kHz and 96 kHz. In my experience, if audio is recorded at 96 kHz, no one can reliably tell the difference between that audio being *played back* at 96 kHz versus 44.1 kHz.

However, I've also found that audible differences can occur between audio *recorded* at 44.1 kHz versus 96 kHz, when processors generate sounds *inside* the computer (as opposed to sounds being recorded via an audio interface). This can happen when using software like virtual instruments and amp simulators.

Without getting too technical, if a signal source like a virtual instrument generates harmonics that interfere with the project's sampling frequency, a certain type of distortion can result. This doesn't happen when recording audio sources feed into an audio interface, because the interface has filters that keep these harmonics from entering the computer. Sound generated *inside* the computer is a different story.

Fortunately, this phenomenon is not a common problem; the issue crops up mostly with older virtual instruments and amp sims. However, it's important to note that claims of a 96 kHz recording sounding better than a 44.1 kHz recording of the same signal may indeed be true, and the difference can be obvious. (Fortunately, there are ways to obtain the benefits of recording at 96 kHz in a 44.1 kHz project; see Appendix A for details.)

Below are some pros and cons of recording at 96 kHz and other sample rates higher than 44.1 kHz.

Pros of Higher Sample Rates

The benefits of using a high sample rate can include the following:

- Under some circumstances, high sample rates may lead to improved fidelity.
- Some people will consider you more "professional" for recording at a higher sample rate, although that's more about perception than reality.
- Higher sample rates can reduce *latency*—if the computer is up to the task.

Latency: A delay that occurs when monitoring audio through the computer. See Chapter 4 for a detailed explanation of latency.

Cons of Higher Sample Rates

The disadvantages of using a high sample can include the following:

- 96 kHz audio requires over twice as much storage space as 44.1 kHz audio.
- Running at 96 kHz stresses the computer more and reduces the potential track count.
- Using a high sample rate may not make any audible difference anyway.

I usually record at 44.1 kHz with 24-bit resolution (see below) so the computer runs smoothly and projects don't need as much space for storage. (Although if a client wants to work at 96 kHz, of course I accommodate them.) However, with instruments and processors that benefit from being recorded at 96 kHz, I use the workaround described in Appendix A.

Bit Resolution

Another aspect of the conversion process is *bit resolution.* This setting specifies the accuracy with which the converter measures the input signal. A good analogy is the accuracy that you can measure with a ruler calibrated in whole inches versus one that delineates to 1/16 of an inch.

CDs use 16-bit resolution, which defines an audio voltage as one of 65,536 levels. However, most audio programs let you record with 24-bit resolution, which provides a much higher level of accuracy—16,777,216 levels. As with higher sample rates, higher bit resolution requires more storage: a 24-bit file is 50% larger than a 16-bit file at the same sample rate. Most recording engineers agree that 24-bit recording is a better choice than 16-bit recording.

In some cases, you may not have a hand in choosing a recording resolution because the interface will default to a fixed resolution. If so, any other options will be grayed out or non-existent. If you can choose a resolution, the most common choices are 16, 24, and 32 bits (Fig. 3.3).

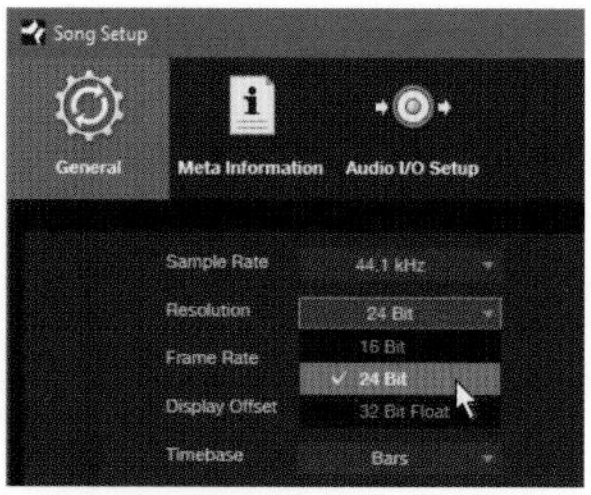

Figure 3.3 Selecting 24-bit resolution in a preferences menu.

Audio Interface Resolution vs. Recording Software Resolution

This is a common source of confusion. The *audio interface resolution* specifies the accuracy with which the converters can capture and play back audio. Resolutions beyond 24 bits are mostly fictional for an interface, because 24 bits already bumps up against the limits of modern technology. You *could* capture or play back audio with more bit resolution, but I feel there would be no audibly relevant advantage.

The *recording software resolution,* also called *process resolution* or *processing resolution,* is different from and usually independent of the audio interface resolution. This resolution is usually much higher—typically 32-bit, 64-bit, and sometimes even more. These resolutions are used because the software has to handle complex mathematical operations. For example, dynamic range changes can involve multiplying and dividing the numbers that represent digital audio. You can easily end up with numbers that require higher resolution—consider this simple example, where you only need 1 digit to represent 2×2. But if you multiple 2×9—which are both single-digit numbers—you now need two digits to represent the result of 18.

So the bottom line is, your audio interface will probably offer 24-bit resolution, though your recording program can operate at a higher resolution. One is not dependent on the other, so you can set your recording program's resolution as high as the software allows. The most common choices are *32-bit floating point* and *64-bit.* In the real world, the results obtained with either option are more or less equivalent.

Tech Talk: File Bit Resolution

This doesn't really have anything to do with the audio interface, but since we're on the subject of bit resolution, your software may also allow you to choose specific resolutions for the files you record or export. For example, if you're doing a live recording and 16-bit audio is sufficient, recording at a 16-bit file resolution creates projects that take up less space. Some people record 24-bit files to match an audio interface's resolution, and it's even possible with some programs to record 64-bit files—although it's probably not worth the extra storage.

Similarly, you can often choose the *render* bit depth. This is the bit resolution of, for example, an exported file. CDs require 16-bit audio, so you would render files intended for CDs with 16-bit resolution. What you choose for the rendering bit depth will also define the bit depth of an audio file you create by converting a virtual instrument output to an audio file.

Key Takeaways

- Most home recording projects use a 44.1 kHz sample rate with 24-bit resolution.
- 96 kHz is considered a high-resolution option; under some circumstances, it can give audibly better sound—but the workaround described in Appendix A can give you essentially 96 kHz performance in 44.1 kHz projects.
- Recording with more than 24 bits of resolution takes up a lot more storage, and it's doubtful anyone would hear a difference in a blind test.
- Recording at 174.6 kHz or 192 kHz stresses computer processing and requires lots of storage. It's mostly used by audiophiles for two-track recordings of live performances and is pretty much overkill for home studios.
- The resolution of your recording software is not the same thing as the resolution of your audio interface. Use at least 32-bit floating point or 64-bit resolution in your recording software.

Chapter 4

Choosing Settings for Latency (Computer-Based Audio Delays)

Musicians expect an instant audible response: Hit a string, blow a note, play a key, and hear a sound. That's why it's important to minimize *latency*—the delay your computer introduces between the time you play or sing a note and when you hear the sound come out of the speakers or headphones.

Explaining this requires getting a bit geeky, but latency can be a tremendous source of irritation and confusion for many people who use computers for recording. Minimizing latency can make the difference between an enjoyable recording experience and a frustrating one, so let's tackle this subject head-on.

If you find this whole subject too confusing, here's a simple rule of thumb: adjust the latency in your recording software's control panel to the lowest possible setting that doesn't affect audio quality or performance.

Latency Basics

Latency occurs in the conversion process from analog to digital and digital to analog, as well as in the computer itself—even the most powerful processor can only do so many millions of calculations per second, and sometimes it can't keep up. To compensate, the computer holds some of the incoming audio samples in a *buffer,* which is like an audio savings account: When the computer is so busy elsewhere that it can't deal with audio, your recording software makes a "withdrawal" from the buffer instead. The larger the buffer, the less likely the computer will run out of audio data when it needs it. But a larger buffer diverts the audio for a longer period of time to keep the buffer filled, which increases latency.

Latency is usually specified in *samples* in an interface's control panel, but it may also be displayed as milliseconds inside the host program. Each millisecond of delay is approximately equal to moving 1 foot away from a sound source. For example, 3 ms of delay is the same amount of delay as moving about 3 feet (about 1 meter) further away from your speakers.

A millisecond, abbreviated ms, equals 1/1,000th of a second.

Most musicians would likely agree that latencies below 5 ms are fine and between 5 to 15 ms are acceptable, but anything much over 15 ms can be annoying. Latencies over 20 to 25 ms can take the fun out of playing and can affect performance.

Unfortunately, latency is a moving target. Projects that have a lot of tracks or loads of plug-ins will tax your computer more than simpler projects, so the computer has to work harder. This requires a bigger audio buffer, which increases the latency.

Adjusting Latency

You adjust the latency in your interface's control panel, but almost all host software will be able to call up this setting so you can adjust latency from within the host (Fig. 4.1).

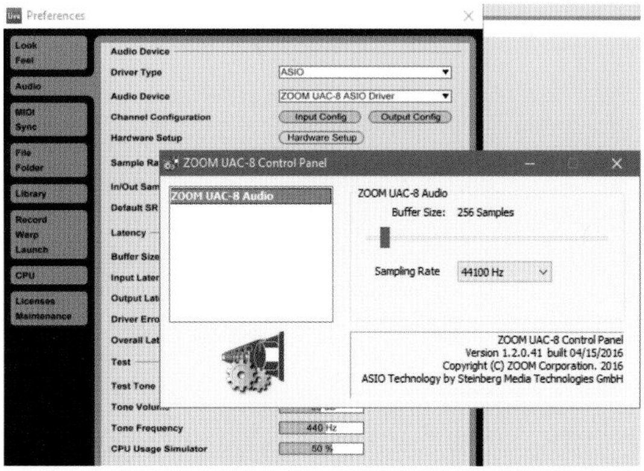

Figure 4.1 Within most recording software, you can open up the interface's control panel to set latency. With Ableton Live, you click on the Hardware Setup button; other software may use the term Control Panel or something similar.

Converting Sample Buffers to Milliseconds of Latency

Although most manufacturers specify the audio buffer in samples, this isn't as intuitive as milliseconds. Some programs include both—you adjust the buffer size in samples, but can also see a readout in milliseconds. If needed, you can convert samples to milliseconds fairly easy, if basic math doesn't make your head explode.

Here are some examples:

- With a 44.1 kHz sampling rate, there are 44,100 samples taken per second. So each sample represents 1/44,100 of a second, or about 0.023 ms. If a sound card's latency is 256 samples, at 44.1 kHz that results in a delay of 256 x 0.023 ms—about 5.8 ms.
- 128 samples of delay at 44.1 kHz would be half as much as 256 samples, or about 2.9 ms.

- At a sample rate of 88.2 kHz, each sample lasts half as long as a sample at 44.1 kHz, so each sample represents about 0.0125 ms. Therefore, a delay of 256 samples at 88.2 kHz would be around 2.9 ms.

Latency Specifications

In some cases, latency specifications are misleading not because manufacturers intend to deceive anyone, but because they can't predict what additional delays a computer will cause. For example, an unspecified delay can be caused by going through a particular USB port, and there may be additional, fixed "safety buffers" that aren't under your control.

As a result, you may find that an interface's theoretical latency is less than what you experience in the real world. Audio interfaces are supposed to report their latency back to the host program, so you can see the amount of latency when making adjustments.

You might assume that a sound card with a listed latency of 5.8 ms is outperforming one with a listed latency of 11.6 ms. But that's not necessarily true, because one program might list the *one-way* latency a signal encounters going into the computer, while another might give the *round-trip* latency—the delay caused by going into, and then returning from, the computer. Today's programs usually give both readings—you might be surprised to see that latency contributed by the sample buffer can end up being only a quarter of the total system latency (Fig. 4.2).

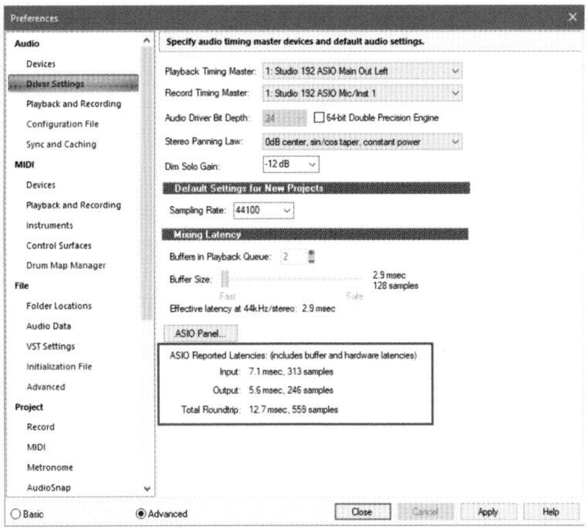

Figure 4.2 Note that although the sample buffer contributes only 2.9 ms of latency, once you include all buffer and hardware latencies, the total "roundtip" latency is 12.7 ms.

To complicate matters, the reported latency readings are not always accurate. Some audio interfaces do not report latency correctly and might be off by several milliseconds. There's no need to get paranoid about this, but be aware that although one audio interface claims a lower latency than another model, you may still sense more of a delay with the "lower latency" audio interface, and it very well might not be lower. Also, some plug-ins increase latency, and this may not be reflected in the latency reading.

So all of this comes back to the question of what latency setting you should use. And the answer remains as stated before: the lowest one that doesn't cause problems!

How to Obtain the Lowest Possible Latency

Before choosing a latency setting, make sure your system is set up to provide the lowest potential latency. You always want the best "feel" when playing and monitoring instruments. To help you get there, here are various ways to minimize latency.

Use a Computer with a Fast Processor

With today's multicore processors, a quality audio interface, and well-written drivers, it's possible to obtain very low latencies. If you have a computer with multiple processors (dual-core, quad-core, etc.), check whether your host recording program supports multi-processor or hyper-threading operation. This increases performance by distributing tasks within the computer over multiple computer cores instead of limiting them to a single core. You may need to look for this option in a preferences or options menu (Fig. 4.3); some programs default to a setting that will work with all computers, which means it's set for the lowest common denominator of a single processor.

Figure 4.3 Assuming your computer supports multiprocessing, check whether your recording software has an option to enable multiprocessing or multiple processing threads.

Check for Driver Updates

Updates can help with anything from squashing bugs to increasing efficiency and reducing latency.

Choose a Modern Audio Interface with a Fast Protocol

For the minimum possible latency, a Thunderbolt 3 interface will outperform other options, at least for now. Recent USB generations like USB 3.1 are faster than previous generations. When it comes to older interfaces, FireWire 800 outperforms the older FireWire 400.... You get the idea.

Adjust Latency Depending on the Project

Even if your computer can theoretically obtain low latencies, in practice lower latencies place a greater burden on the processor. Because most audio interfaces provide a choice of latency settings, you can choose the "sweet spot" between low latency and computer performance. If your computer has to record only a few tracks, then you can probably adjust your settings for really low latency. But if you're running a complex digital audio recording program and playing back lots of tracks with virtual instruments, you may need to set the latency higher.

Some musicians make it a practice to record parts that require a lot of computer power—like guitars using amp sims, or virtual keyboards—early on in a project. This lets them enjoy the lowest possible latency settings. Later on when mixing, where low latency isn't as important because they're no longer recording instruments in real time, they'll increase the latency setting.

Freeze Virtual Instrument Parts

Most recording software has an option to *freeze* virtual instrument parts. This renders the instrument part as audio, which requires less CPU power to play than the virtual instrument. If you need to make changes to a frozen instrument part, you can unfreeze it, make the required changes, and then refreeze it.

In a project with lots of virtual instruments (requiring high sample buffers), freezing the parts can allow for reduced latency. Consult your recording software's documentation for details on how to freeze and unfreeze instrument tracks.

Select the Right Driver Protocol for Your Sound Card and Program

On the Mac, it's easy: Choose Core Audio. On Windows, an audio interface might offer several possible driver protocols, such as ASIO, DirectX, MME, emulated ASIO, etc. If an audio interface includes an ASIO driver written specifically for the device (it should), that's the one to use. Typically, it will include the manufacturer's name. Do not use an "emulated" driver, as it will run much more slowly.

Monitor through Headphones

If you're listening to monitor speakers and you're sitting around 3 to 4 feet away from them, you've just added another 3 to 4 milliseconds of delay. Monitoring through headphones will remove that latency, leaving only the latency caused by using the audio interface and computer.

Record Projects at Higher Sample Rates

Because a higher sample rate causes the computer to process the audio more rapidly, it fills and empties the buffer faster. However, whether the computer can process the audio more rapidly is not always a given.

Use Zero-Latency Direct Monitoring

An audio interface feature called *zero-latency monitoring* or *direct monitoring* reduces latency to virtually nothing, so what you hear as you monitor is essentially in real time. However, interfaces do this by monitoring the signal at the audio interface's audio input, thus bypassing the computer completely—it's as if you used a Y-cable, and sent one feed to the audio interface and the other one to your monitor speakers or headphones. Therefore if your audio playback requires software running inside the computer, like a virtual instrument, amp sim, or crucial plug-in, zero-latency monitoring will not be appropriate as the computer processing will not be included. On the other hand if you're a singer who just wants to hear yourself in the headphones, zero-latency monitoring is very useful.

Key Takeaways

- Lower latencies give the best experience when using virtual instruments and plug-ins—use the lowest sample buffer possible, consistent with the audio not breaking up or distorting.
- If possible, record CPU-heavy parts early in a project so you can use lower latency settings than if you record those parts in a project that already has a lot going on.
- Some plug-ins increase latency. Avoid using them while tracking, or substitute something similar that requires less CPU power until mixdown.
- The interface technology you use influences latency. Thunderbolt interfaces have the least latency.
- Several workarounds are available to reduce latency; see the section "How to Obtain the Lowest Possible Latency" in this chapter.

Chapter 5

Digital Audio Connections

Manufacturers typically define an interface's inputs and outputs by listing a spec with two numbers. For example an 8 × 2 interface has 8 inputs and 2 outputs. Sometimes these numbers don't seem to match the physical connections on the interface. You might wonder, for example, why an interface with eight analog inputs lists 18 or more inputs. The listed spec includes the total number of inputs and outputs that are available. Analog signals aren't the only way to get audio from one piece of gear to another—it's also possible to transfer digital audio. So, many audio interfaces include connections for digital inputs and outputs.

Some equipment will have both analog and digital outputs. Where available, this allows sending digital output signals from the gear to an audio interface's digital input, which will give higher quality than sending and receiving analog signals, because (among other things) the direct digital connection avoids conversion back and forth between digital and analog. It's likely you won't hear a significant difference, but I'm a big believer in taking advantage of any improvements you can, because those improvements add up over multiple tracks.

The sections below describe the most common digital protocols used in audio recording.

S/PDIF (Sony/Philips Digital Interface)

This is a very common digital connection on audio interfaces because S/PDIF outputs are also found in some consumer electronics. S/PDIF carries a stereo digital audio signal and typically uses either RCA (phono) electrical coaxial connectors (Fig. 5.1) or TOSLINK fiber-optic connectors for input and output connections.

Trivia time! TOSLINK is short for Toshiba Link and is a registered trademark of Toshiba. The generic name is EIAJ Optical.

Figure 5.1 Coaxial S/PDIF connections typically use RCA phono connectors, although sometimes in professional gear, XLR connectors get the nod.

S/PDIF can also use XLR connectors, which are preferred for AES/EBU digital audio signals (described below). However because the levels of S/PDIF and AES/EBU are different, there's usually a switch to select between the two types of digital audio signals.

Optical connectors included for S/PDIF may also work with digital audio signals based on the ADAT optical interface described below; however this is not always certain. Although most S/PDIF connections are coaxial, some interfaces use optical TOSLINK connectors for both S/PDIF and ADAT (Fig. 5.2).

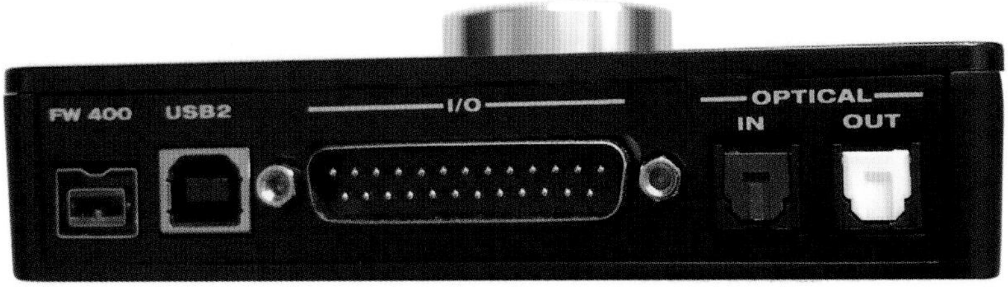

Figure 5.2 MOTU's Track 16 interface's TOSLINK optical input and output connectors are compatible with S/PDIF and also the ADAT multichannel optical interface. The 25-pin connector mates with a breakout cable that provides two line inputs, guitar input, two main outs, two line outs, two XLR mic inputs, MIDI in, and MIDI out. (This interface also has the unusual feature of offering both FireWire 400 and USB 2.0 ports.)

AES/EBU (Audio Engineering Society/European Broadcast Union)

This is another two-channel protocol. AES/EBU is very similar to S/PDIF, but it uses balanced, XLR-style connectors and is more of a professional format (Fig. 5.3). Some audio interfaces offer both AES/EBU and S/PDIF connectors.

Figure 5.3 Input and output XLR connectors for AES/EBU digital audio connections.

ADAT (Alesis Digital Audio Tape) Optical Interface

This 8-channel protocol uses TOSLINK connectors and sends/receives signals over a fiber-optic cable. While originally used with the Alesis ADAT recorders, over the years it has become a *de facto* multichannel standard—mixers, hard disk recorders, and other devices often include an ADAT connector (Fig. 5.4).

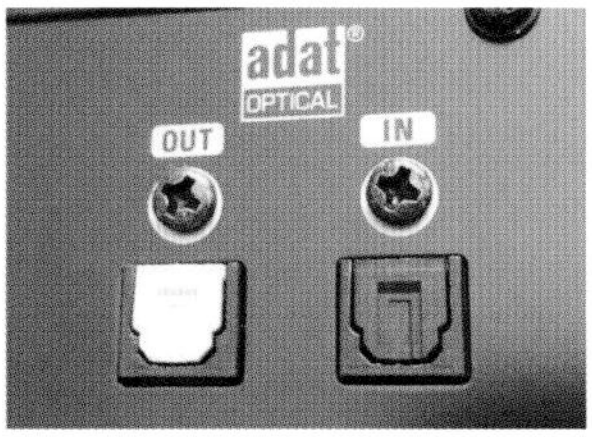

Figure 5.4 Although ADAT digital tape recorders are a rarity, the ADAT optical connection remains viable.

Perhaps the most important current application of an audio interface's ADAT input is to expand an audio interface's number of input channels. Several 8-channel mic preamps are available with an ADAT output. You can connect the ADAT out from such a unit to the ADAT input on your audio interface, and *voilà*—eight extra mic inputs. Compact, desktop-type interfaces with limited inputs often include an ADAT input to allow for easy expansion.

ADAT S/MUX

This format is similar to the standard ADAT protocol but addresses the limitation of the original spec that could not handle sample rates above 48 kHz. The S/MUX protocol *multiplexes* (splits) a higher sample rate signal (88.2 kHz or 96 kHz) between two digital audio streams by sending one sample over one stream and the next over the other. So, even though the digital audio streams run at 44.1 or 48 kHz, when demultiplexed, the result is a higher sample rate. The drawback is that ADAT S/MUX can carry only half as many channels of audio as the original ADAT spec (Fig. 5.5).

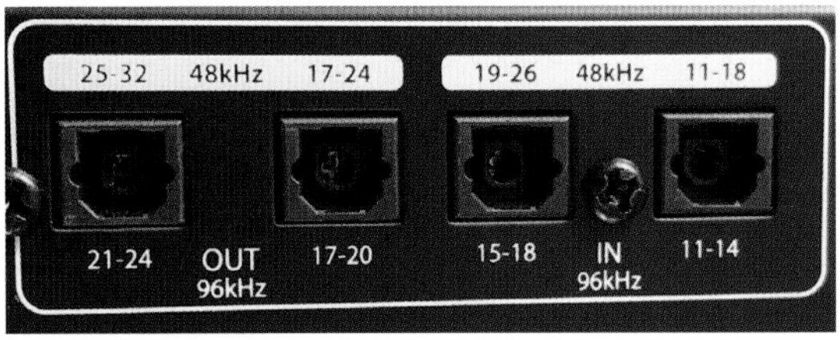

Figure 5.5 A pair of ADAT connectors can create the ADAT S/MUX interface for sample rates higher than the original 44.1 kHz/48 kHz options. Labels on the rear panel of this PreSonus interface show the 16 channel assignments for standard 44.1/48 kHz ADAT connections (the label above the TOSLINK connectors) and the eight channel assignments for 88.2/96 kHz S/MUX assignments (below).

ADAT S/MUX IV

Another more rare alternative, S/MUX IV, accommodates two channels of 192 kHz/24-bit audio over a standard ADAT optical cable.

MADI (Multichannel Audio Digital Interface)

Originally designed to transfer 56 channels of audio over coaxial cables, MADI is used by higher-end consoles and interfaces. An enhanced version, Extended MADI (also called MADI-X) provides up to 64 channels at 48 kHz and 32 channels at 96 kHz, and can also use optical or Cat 5 cables.

> Cat 5 cables (or Category 5 cables) are the kind commonly used for Ethernet connections. Modern variants include Cat 5e, Cat 6, and Cat 6A, which are backwards compatible with Cat 5.

RME is one company that makes a range of MADI interfaces (Fig. 5.6), but so far MADI is rarely used in home and project studios.

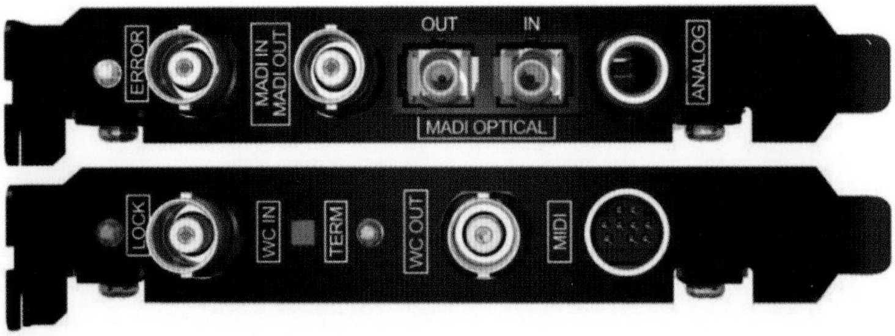

Figure 5.6 The MADI interface can transfer up to 64 audio channels over wire or optical connections. The RME HDSPe interface includes both options.

MIDI (Musical Instrument Digital Interface)

MIDI is a hardware/software protocol designed primarily for musical instruments. The MIDI software language describes musical parameters like notes, dynamics, tempo, pitch bend, modulation, and the like. A MIDI-compatible keyboard generates data that represents these kinds of performance gestures, and since it's all data, a computer can easily record it. Then, the computer can play back this data into a MIDI-compatible tone generator to re-create the performance—very much like a digital player piano.

A good analogy is a computer QWERTY keyboard. When you type, the keyboard sends out data that represents the letters, numbers, and symbols you type. The computer can record these keystrokes as words, then output them to a printer.

The original MIDI protocol specified a five-pin *DIN connector*, which is what you'll still find on audio interfaces for MIDI in and MIDI out. These transfer data into the computer and send out computer data to a MIDI-savvy sound generator, respectively (Fig. 5.7).

DIN connector: This style of connectors takes its name from the Deutsche Institut für Normung, a German standards organization that standardized the connection type in the 1970s.

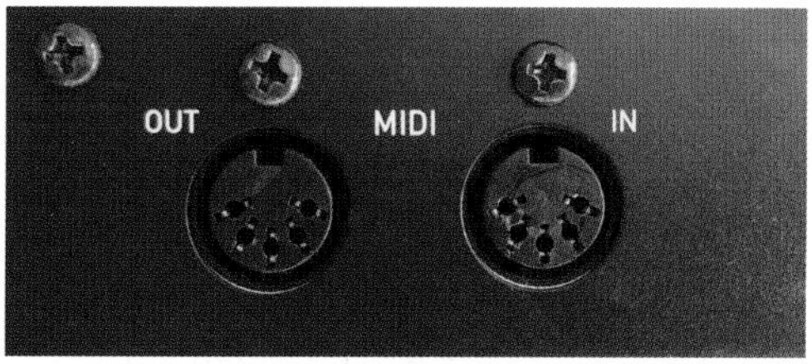

Figure 5.7 Five-pin MIDI out and MIDI in jacks were the original MIDI connectors, and are still in use. However, a lot of today's MIDI gear sends and receives data over USB instead.

MIDI data can also be carried over USB, and many newer MIDI devices connect to computers with USB rather than 5-pin DIN-compatible cables; some MIDI devices have both options. Older MIDI gear uses only 5-pin DIN connectors, so if you use any you'll want these connectors on your audio interface.

If your audio interface doesn't have 5-pin DIN connectors but you later find you need them, it's not a problem. MIDI interfaces are available that connect with a computer's USB port and provide MIDI in and out with 5-pin DIN connectors. Some of these interfaces are simply cables, which makes them very convenient if you have an instrument or two that you need to connect to your computer via USB (Fig. 5.8).

Figure 5.8 ESI's MIDIMATE eX provides DIN-to-USB and USB-to-DIN conversion.

Basic Clocking

As mentioned, digital systems are based on particular sample rates. A master clock (also called *system clock*) determines the sample rate frequency. For best results, this clock should be as accurate and stable as possible. Furthermore, all digital audio devices should sync to this clock. Think of it this way: The system clock is the drummer, and all the other "musicians" (digital devices) get their timing cues from the drummer. For the best rhythm, you want the drummer to be "in the pocket," and you want the other musicians to be able to follow accurately. It's similar with digital systems.

Your host software's preferences will specify a clock source. In most cases, clocking is handled transparently and you don't need to think about it. This is because the clock source will almost always default to Internal, which means the software is receiving its clock signal from the connected audio interface (after all, the interface is already working at the desired sample rate).

Clocking becomes an issue only as more gear is involved; the general rule of thumb is to set the clock to the digital output of a device feeding into your interface's digital input. For example, if a unit with a S/PDIF digital output connects to an audio interface's S/PDIF digital input, the two devices need to synchronize to avoid issues with audio going out of sync. Sync problems are heard as clicks or dropouts. Because the S/PDIF signal is outputting a clock signal, you can synchronize the interface to it by specifying S/PDIF as the clock source. If the interface is later used to receive a signal from an ADAT output, you would then specify ADAT as the clock source.

Word Clock

The above is fine if you deal with only two pieces of gear, but when you have multiple pieces of gear that all have to synchronize to each other, *word clock* comes into play. This is an advanced topic and you will likely not need to be concerned about it unless your setup requires synchronizing multiple pieces of gear. So, you can probably skip the following information—but it's here if you need it (or are naturally inquisitive).

In higher-end digital audio systems, you'll often find a stable master clock that outputs its clock signal over a *BNC connector*. This clock signal typically gets daisy-chained—in other words, the master clock output feeds device A's clock input, device A's clock output feeds device B's clock input, device B's clock output feeds device C's clock input, and so on (Fig. 5.9).

BNC connector: A miniature quick connect-style linkage featuring two bayonet lugs on the female receptacle. BNC stands for Bayonet Neill-Concelman, so-named after the connector's shape and designers.

Figure 5.9 The BNC IN connector (right) receives word clock, while the OUT connector (left) passes the word clock signal along to the next word clock input.

The last input is *terminated*, meaning that it includes circuitry that tells the clock signal, "This is the end of the line." (For the technically-minded, termination prevents the clock signal from reflecting back into the line.) Most digital audio devices intended for use with external clocks will have a termination switch at the input in case it's the last unit in the timing chain. (Incidentally, unlike AES/EBU, S/PDIF, and ADAT, the MADI interface mentioned above does not include a clock signal. It's necessary to use an external clock generator with MADI.)

In some digital audio systems, instead of using daisy-chaining, the clock goes through a distribution amplifier that feeds each input individually.

Fun stuff, eh?

Hey! *Wake up*!!!

Networked Audio

Two common, non-proprietary methods are available to network audio: Dante (Digital Audio Network Through Ethernet) and AVB (Audio Video Bridging). These are intended primarily for commercial and professional applications, where many audio channels need to be sent to several locations, and/or over long distances. As such, this topic is not really that relevant to home studios unless you have a spread out arrangement—for example, a setup where you track at one end of a house and you mix in a converted upstairs bedroom. Even then, more conventional approaches will often do the job.

A deep dive into networked audio is beyond the scope of this book. However, Mitch Gallagher has written an excellent article for Sweetwater entitled, "Audio Networking Explained." This article can be found online at https://www.sweetwater.com/insync/audio-networking-explained/.

Key Takeaways

- A common use for digital connections is to expand an existing interface with more inputs by feeding an expander module with an ADAT output to an audio interface with an ADAT input.
- S/PDIF has been available on some consumer gear, but there's not a lot of need for stereo digital audio transfers any more—the days of transferring mixes to Digital Audio Tape or recordable CDs are mostly behind us in our streaming-oriented world.
- For clocking, most of the time selecting *Internal* in your audio interface's preferences is all that's needed. If one device is feeding another device through a digital audio connection, the receiving device should sync to the transmitting device (typically ADAT or S/PDIF clock source).
- If a setup involves multiple pieces of digital gear that work together with digital connections, they need to be synchronized via word clock, if they're not networked.

Chapter 6

Microphone Preamps

Microphones generate very low-level signals, so they require high-gain preamps to amplify the audio up to line level. An audio interface's sound quality depends on the quality of its preamps. Fortunately, technology has progressed to where even relatively low-cost audio interfaces offer preamp quality that equals the quality of high-end preamps not that long ago. Marketing departments will wax poetic about their "clean, warm, transparent" preamps, but the reality is that the difference between most modern preamps in a given price range is more like the difference between "excellent" and "even somewhat more excellent."

Although most interfaces will have mic preamps, it's also possible to buy stand-alone or specialty mic preamps and connect their outputs to an audio interface's line inputs. The line inputs may bypass the internal mic preamps or have their gain adjusted by the preamp gain controls anyway.

Preamp Gain

The amount of gain available on a preamp is specified in decibels. The higher this number, the greater the available gain—e.g., a preamp with 60 dB of gain can provide more amplification than a preamp with 50 dB of gain. When close-miking instruments, just about any amount of gain will be enough. You'll need more gain for mics recording room ambience, whispers, quiet acoustic instruments, and the like.

For most microphones, 50 to 60 dB of gain is sufficient. Ribbon microphones, once rare but now becoming more common thanks to improvements in manufacturing techniques, often require more gain—as much as 70 dB for quiet signals. Few audio interfaces can provide that amount of gain (check the specs), so you can either turn up the preamp gain as much as possible and hope that the signal-to-noise ratio is good enough; or use an external, high-gain preamp and plug its output into an audio interface's line input (Fig. 6.1).

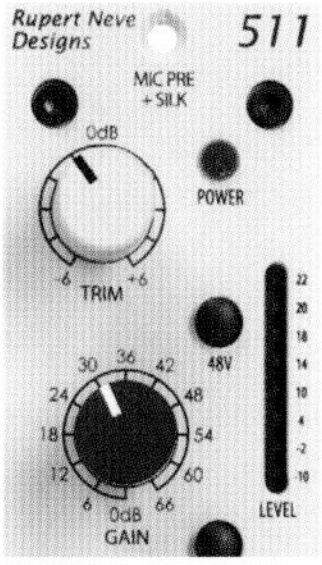

Figure 6.1 The Rupert Neve Designs 511 preamp provides from 6 to 66 dB of gain, so it can handle just about any miking situation. An additional Trim control can add six more decibels, if needed.

Preamp Design Classes

Preamp circuit design has various classifications. The two most common types are Class A and Class AB. One class isn't inherently better than the other; a well-designed Class AB preamp with quality components will sound better than a poorly designed Class A preamp with cheap components. And of course, how a preamp sounds to *you* is ultimately what matters—your personal preference may favor a totally clean and neutral preamp or one with some character.

That said, all things being equal, a Class A preamp will deliver more accurate amplification than a Class AB type, which is designed for efficiency (that's why it's commonly used for power amps). The tradeoff with Class A preamps is that, because they require more current, it may not be possible to bus-power an audio interface with many Class A mic preamps. That means you'll need to use an AC adapter.

For many years, a person's go-to preamp could easily be one with a certain characteristic sound instead of one that was totally neutral. However, as digital technology has become more refined and plug-ins have developed the ability to reproduce ever-more nuanced sonic characteristics, the trend is to want neutral preamps that add no character of their own. That way, any character will be added only by plug-ins based on choices made by the engineer—it's analogous to why painters want a white canvas.

Understanding Mic Preamp and Audio Interface Specifications

The most important audio interface specification involves its mic preamps and associated circuitry. Although audio interfaces usually provide specs for the interface as a whole (and not the mic preamps per se), the mic preamps are the most crucial factor in determining these specs.

Spoiler alert: Most manufacturers' spec sheets are pretty meaningless because they don't provide the conditions under which the tests were conducted. For example, the results for just about any characteristic will vary depending on whether the preamp gain is set to zero or set to maximum. Results can also depend on the reference level—for a given amount of noise, if you're measuring signal-to-noise ratio by comparing the maximum possible output level to the residual noise, that spec will seem much better than a measurement taken with a specific reference level that's lower than the maximum possible output.

With that in mind, let's look at the meaning of five important characteristics that are commonly listed in spec sheets: frequency response, signal-to-noise ratio, total harmonic distortion, intermodulation distortion, and crosstalk. I use a software analyzer to measure these specs myself, to ensure that the conditions under which the measurements are taken remain the same. This way I can compare apples to apples and get a better idea of how units compare as opposed to looking at specs provided by the manufacturer. The specs cited in the following discussions are all taken from TASCAM's US-4×4 interface, which has very respectable specs for a unit in its price range.

Frequency Response

Ideally, audio gear designed for maximum accuracy should reproduce all audible frequencies equally—bass shouldn't be louder than treble, or vice-versa. A frequency response graph shows what happens if you feed test frequencies with the same level into a device's input and then measure the output for variations. You want a response that's flat (even) from 20 Hz to 20 kHz, because that's the audible range for humans with good hearing. It becomes harder to reproduce extremely high or extremely low frequencies. Fig. 6.2 shows a typical frequency response graph.

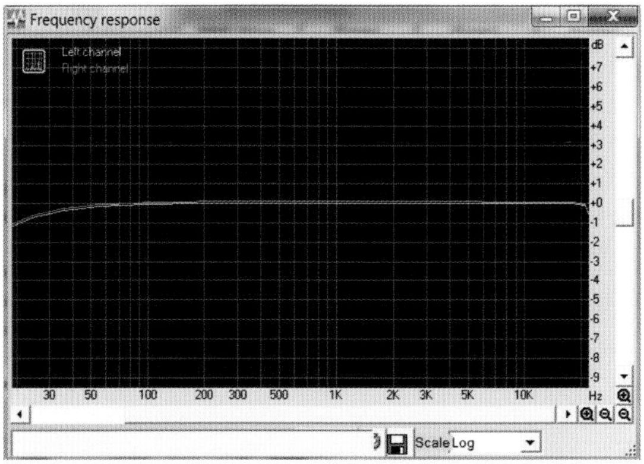

Figure 6.2 This graph shows that the output has essentially no response deviations from 50 Hz to 20 kHz, and is down one dB at 20 Hz.

As with most audio interfaces, the response in Fig. 6.2 drops off even further below 20 Hz; this is deliberate, because there's no need to reproduce signals we can't really hear. This graph shows that the interface has a high degree of accuracy when reproducing signals at audible frequencies.

A manufacturer's specification will give a frequency range and amount of response deviation. For example, the spec for the above interface could be expressed as –1 dB, +0 dB from 20 Hz to 20 kHz, or ±0.5 dB from 20 Hz to 20 kHz.

Signal-to-Noise Ratio

All electronic circuits generate some noise, but of course you want the lowest possible noise level. Noise increases as you turn up the gain. For example, here's the US-4 × 4's mic preamp noise with the volume turned up one-fifth of the way (Fig. 6.3).

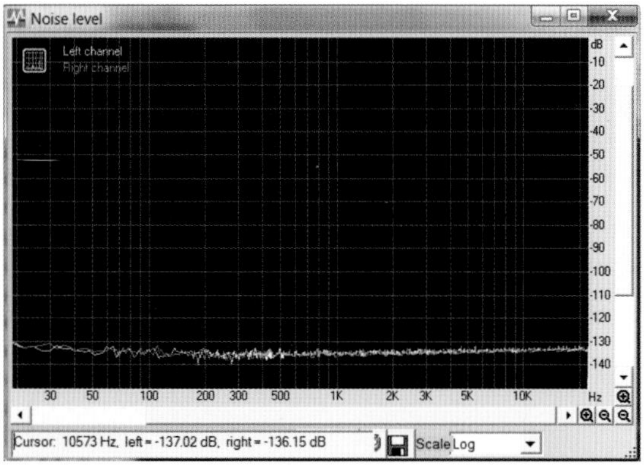

Figure 6.3 The noise is less than –130 dB. In other words, compared to a signal at full level, the noise is over 130 dB softer.

This graph shows that the signal-to-noise ratio (the ratio of the full-level signal to the noise) is 130 dB, which is extraordinarily quiet. But what happens if we increase the mic preamp gain two-thirds of the way—about right for recording a really quiet vocalist? Here the noise level has increased, but it is still under -110 dB (Fig. 6.4).

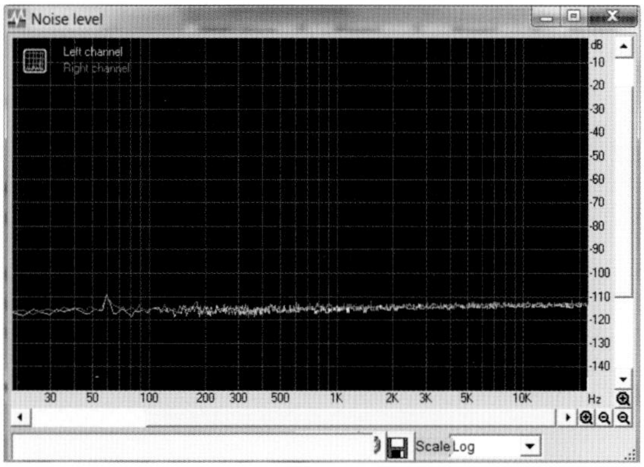

Figure 6.4 Even with the gain turned up, this is still extremely quiet. Consider that Compact Discs can at best reproduce levels no lower than about –90 dB. Essentially, the noise is so low a CD can't even reproduce it.

These two graphs also hint that specs have the potential to mislead rather than enlighten. Company A might use a graph like the first one for their marketing, while Company B might choose a more real-world graph like the second one. Company A could imply their mic preamp is quieter—"Just look at the specs!" So it's important to know the conditions under which measurements are taken, and of course, these usually aren't given (and to be fair, most people wouldn't know how to interpret them anyway). In this chapter, most of the measurements I'm showing reflect both ideal and real-world conditions. However these particular mic preamps hold up well under that kind of scrutiny.

Noise specs are given in different ways, so make sure you compare the same kind of spec. The noise spec may be given as EIN (Equivalent Input Noise), which represents the amount of noise a mic preamp adds to a microphone's signal. The lower the number, the better. These numbers are very small, which make the preamps look good. Another option is signal-to-noise ratio, which compares the ratio of a specific signal level (usually for a given, low level of distortion) to the residual noise. For example, a signal-to-noise ratio of 90 dB means the signal is 90 dB louder than the noise.

The noise spec may have a particular "weighting." This means that the spec excludes noise that isn't relevant to our hearing. Another option, dynamic range, is the ratio of the maximum possible signal to the residual noise.

Total Harmonic Distortion

As all guitarists know, distorting a signal adds harmonics—and we love them! But just as all circuits generate some degree of hiss, they also generate some degree of undesirable distortion. Total Harmonic Distortion graphs show the level of the harmonics generated by distortion.

This test feeds in a 1 kHz signal at maximum level. In theory, the output should show a spike only at that 1 kHz signal. Any other signals represent distortion (Fig. 6.5).

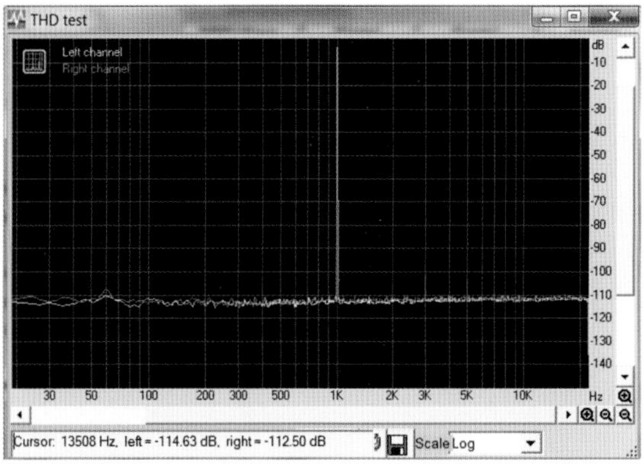

Figure 6.5 The graph shows a little bit of 3rd harmonic distortion at 3 kHz and 5th harmonic distortion at 5 kHz. Both are below –100 dB.

Although there's additional harmonic distortion not visible in the graph, it's so low it's below the noise. If we turn the gain all the way down to minimize noise, you can then see these additional distortion components (Fig. 6.6).

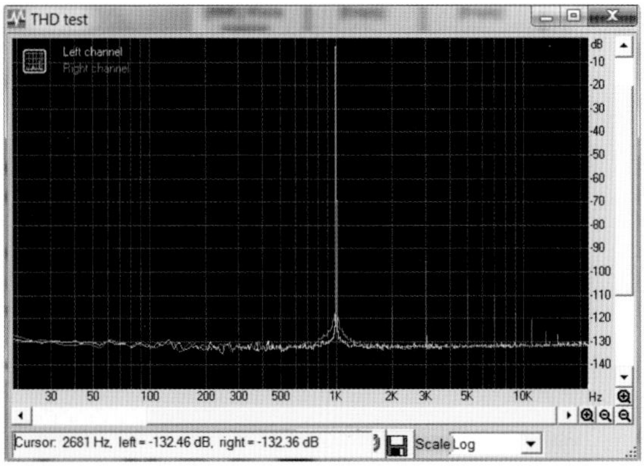

Figure 6.6 With the preamp gain all the way down, the distortion components at 7 kHz and above become visible above the noise.

Intermodulation Distortion

This type of distortion occurs when two signals interact with each other in such a way that they produce artifacts. These artifacts also represent distortion, and many people consider intermodulation distortion more objectionable than harmonic distortion. The standard version of this test feeds in two signals at maximum level, one at 60 Hz and one at 7 kHz. Any output signals other than these two frequencies represent distortion (Fig. 6.7).

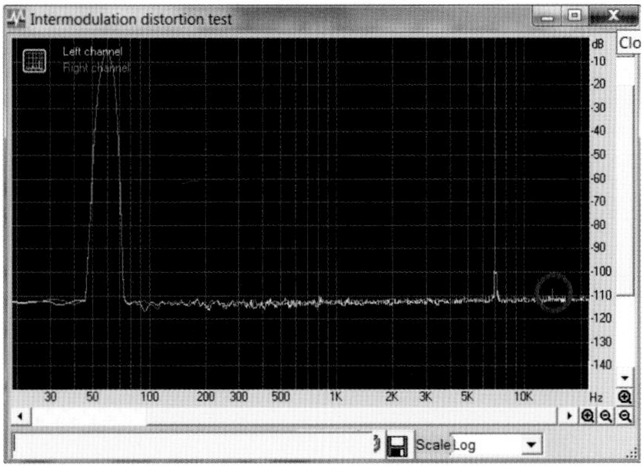

Figure 6.7 The intermodulation distortion is so low it's very difficult to see on this graph. But if you look closely, you'll see a little spike (circled in red) poking up from the noise.

Reducing preamp gain reduces the noise floor, which reveals more distortion artifacts (Fig. 6.8).

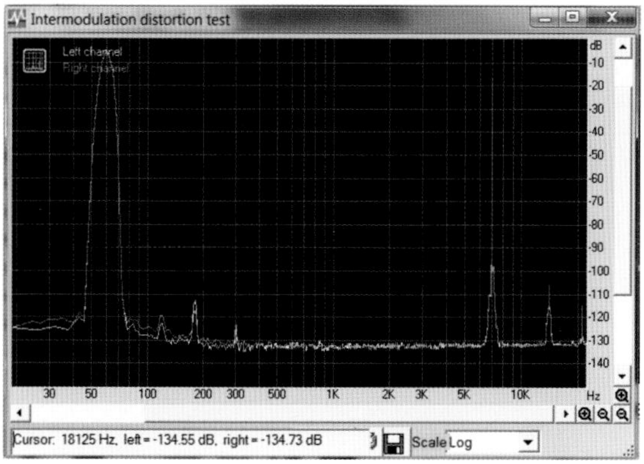

Figure 6.8 With no gain, you can see four low-level amounts of distortion below –110 dB.

Crosstalk

Crosstalk occurs when one channel picks up some of the signal from another channel. This happens because some circuit elements radiate signals, while other elements can pick up those signals. Careful mechanical design and signal isolation can reduce crosstalk, but you can't get rid of it entirely. Crosstalk (Fig. 6.9) is more likely with high gain, high frequencies, and sometimes, low frequencies.

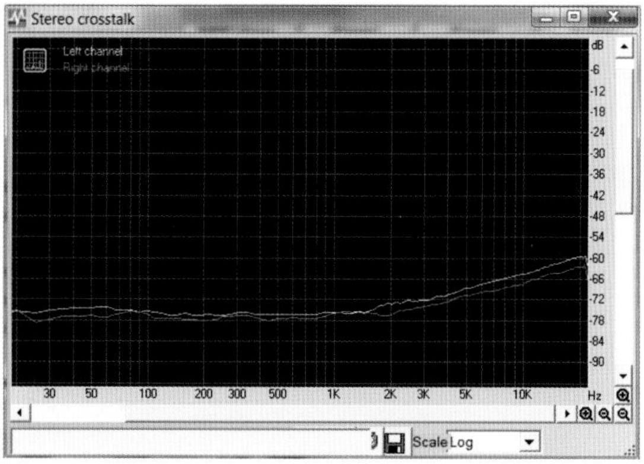

Figure 6.9 This crosstalk measurement, taken with a high-gain setting, is comparable to similarly priced devices.

When you turn down the gain to a level suitable for recording an instrument like steel-string guitar, crosstalk goes way down (Fig. 6.10).

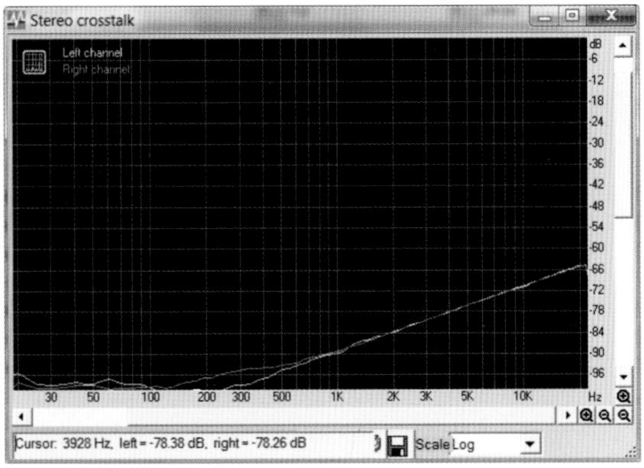

Figure 6.10 Reducing the gain helps reduce crosstalk considerably.

This helps illustrate an important point: Specs can be a "snapshot" of a gear's particular state, or be represented by a range of numbers. Always use specs as a guide, not a judge—and if you'd like to know more about why the specs you read are pretty much meaningless, Rane has published a very deep explanation of specifications at http://www.rane.com/note145.html.

Key Takeaways

Many would argue with the following, so take my opinions with a grain of salt (or pepper, if you like spicier stories).

- Most specs from manufacturers are fairly meaningless because they don't all use the same references and tests. However, …
- Mic preamps in the same general price range have more similarities than differences.
- Class A mic preamps can theoretically give better performance than Class AB types, but they draw more current.
- Sometimes 50 dB of gain isn't quite enough. Gain of 60 dB is almost always sufficient unless you're using ribbon mics, which often need 70 dB of gain.

Chapter 7

Other Audio Interface Features

Audio interface manufacturers are always looking for unique features they can add to their products to help them get ahead of the competition. Here are some common ones.

Input Level Metering

Virtually all audio interfaces provide input level metering, but the sophistication varies. There's even some question as to how necessary these LED meters are—after all, you may not be able to adjust the input level in the recording software itself, so you'll adjust the audio interface's input levels based on the recording software's meters or a mixer application that comes with the interface. If the meters show a red clipping indicator, you would simply turn down the associated input level control on your audio interface.

However, audio interfaces can also serve as mixers for live performance (as described later in this chapter). In this case, metering becomes much more important because you wouldn't know what's happening with the signal unless you had a computer or other means to display routing and levels.

The simplest indicator is a single LED or other indicator to show *clipping*, meaning that the input signal has exceeded the available headroom. A clipping indicator is better than nothing, because it helps you avoid distortion. Unfortunately, it's not always obvious what the clip light means. It may light only if the signal has actually gone into distortion, in which case it's too late to do anything about the problem. Or it may light a few dB below clipping, which may alert you to turn things down before problems occur. To find out what the clip indicator means, feed a steady tone (like a synthesizer note set for a sine or triangle wave) into the interface input, slowly turn up the level until the clipping indicator lights, and then observe the recorded waveform to see if it's distorted or not.

Better input metering options will include more than one LED, or a bi-color LED to indicate green for signal present and red for clipping. The signal present LED is helpful if you're not seeing audio going to your recording program. That way you can determine whether the problem is the signal going to the interface, or perhaps an incorrect setting inside the software itself (like forgetting to arm recording).

The best input metering provides multi-stage meters with three or more indicators to show a series of levels (Fig. 7.1). An interface may even include an analog meter (the mechanical kind with a pointer) to give more of a vintage vibe.

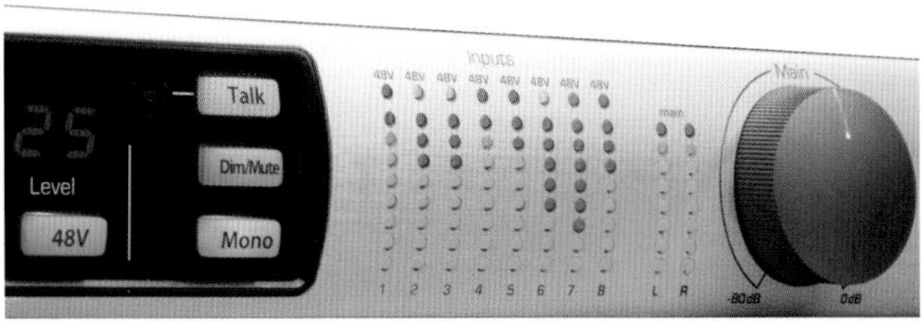

Figure 7.1 The PreSonus Studio 192 has eight-step meters for all eight analog input channels.

Onboard Effects

Some interfaces include onboard digital signal processing (DSP) to provide effects like compression, EQ, and reverb. Whether onboard DSP is useful or not depends on how you plan to use it.

For example, it may seem like compression is a good idea because it can prevent short overloads from making it through the interface, and therefore minimize the odds of recording a distorted signal (especially for a live recording, where you don't get a second chance). Yet with virtually all interfaces, this is not possible because the compressors come *after* the A/D converters, not before—where they could indeed prevent excessive levels from making it into the converters.

Some might disagree, but I think the main advantage of DSP is making performers happier. If reverb is available, you can send it to a singer's headphones so they can hear some reverb as they sing. But you don't need to record the reverb in your recording software, so you can decide what kind of reverb to add during mixdown. Similarly, a bass player might want to monitor a compressed bass signal, because it likely will be compressed during mixdown anyway. This feature allows the bassist to play while taking the eventual compression into account.

Universal Audio's Apollo Interfaces

Universal Audio takes a novel approach to onboard DSP with their Apollo interfaces. The included hardware-based effects can process your signal as it's being recorded into your host program and provide monitoring with near-zero latency so performers can hear exactly what something will sound like when played back through those plug-ins—but without having to deal with latency. In other words, it works the same way as zero-latency monitoring, but with the bonus of including effects (Fig. 7.2).

Figure 7.2 A limiter, amp sim, and room simulator are processing guitar in the Apollo mixer application. You can monitor with near–zero latency through the effects and either record the signal with effects to your host software or record the unprocessed signal for later processing during mixdown. In this screen shot, the effects are being recorded into the host software.

In general, engineers frown on recording with effects unless it's a crucial part of someone's sound. However, Universal Audio is known for (among other attributes) emulating the character of various legendary consoles. By recording through these effects, you can have someone else work on the file in a different studio who may not have these effects, yet still have the desired sonic character "baked into" the audio. Also, some musicians prefer to commit to a sound early on rather than waiting until mixdown to decide what the sound should be.

Mic Preamp Low Cut and Phase Switches

Low-cut switches can reduce pops and other low-frequency artifacts you'll just want to get rid of anyway while recording. In some cases these switches are included in the hardware preamps (as with the Focusrite Clarett 8PreX; see Fig. 7.3) so you can minimize pops and booms prior to the A/D converters. This allows for more headroom and may help prevent distortion.

Figure 7.3 The Focusrite Clarett 8PreX has front-panel, hardware switches for low cut (to the right of the +48V phantom power switch) and phase (to the right of the low-cut switch).

On the other hand, if the low-cut switch is software-controlled, it may be in the interface's internal DSP and therefore won't condition the signal prior to the A/D converter.

Phase switches flip the phase (also called polarity) of the mic signal. Although theoretically this shouldn't make a difference with mono signals, some people make sure that the polarity is the same as the source—for example when miking a kick drum, pushing on the kick drum beater pushes air outward from the kick drum head. Check that the speaker cone pushes outward when the kick drum hits. If not, consider reversing the phase.

A more common issue occurs when using two (or more) mics on the same signal source. If mixing the signals together in mono produces a weaker or thinner sound, reverse the phase of one of the mics to see if that improves the sound. Note that you can typically reverse phase in your host software as well, but recording with the proper phase relationship means one less item to consider when mixing.

Digital Mixer Capabilities

Because audio interfaces have inputs, outputs, and the ability to alter levels, they can serve as digital mixers. Some devices handle this by being essentially mixers with interface capabilities, and including all the controls that implies (Fig. 7.4).

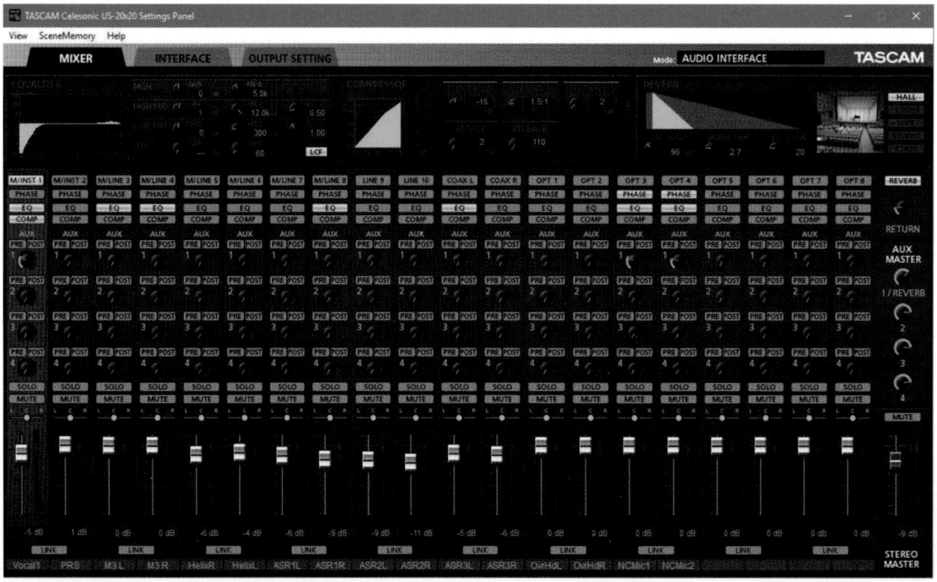

Figure 7.4 The mixer page from TASCAM's US-20x20 interface application shows that yes, it really is a mixer.

Using a rack-mount interface as a mixer is not as ergonomic as using a traditional mixer, but such a device can still be useful as a keyboard mixer, a mixer for a singer/songwriter setup with a couple mics and guitars, and the like. The main disadvantage is that there aren't any front-panel controls you can use to adjust settings easily. Instead, you'll need to hook the interface up to a computer and use the associated application to set levels and routing. However, sometimes all you really need is a "set and forget" mixer, and there are usually enough front-panel controls to get the job done.

Software Applications

Virtually all audio interfaces come with companion applications for setting levels and routing. These range from the simple to the bewildering, and can include sophisticated functions like looping audio from an output back to an input, incorporating plug-ins for processing, and more (Fig. 7.5).

Figure 7.5 Roland's companion application for the Octa-Capture offers many standard mixer features, including compression.

These applications can often store presets. So for example, if you're recording vocals, you can save a preset with the desired level setting for the vocal mic and all other inputs turned off so they don't contribute noise. Often it's possible to mute or solo individual interface channels, which is handy for diagnostic purposes (e.g., to find out which ones are contributing that weird hum you keep hearing). Granted you can also do this in your recording software, but sometimes you might not want to disturb those settings.

Remote Control Programs

Some audio interfaces, being more or less like mixers, offer tablet or smartphone remote control applications (Fig. 7.6). While this may seem redundant—after all, you can adjust the parameters on your main computer or on the unit itself—there are valid reasons for using remote control.

For example, if you're a solo artist, you can set levels from where the mic is located instead of having to constantly go back and forth between your mic and the computer.

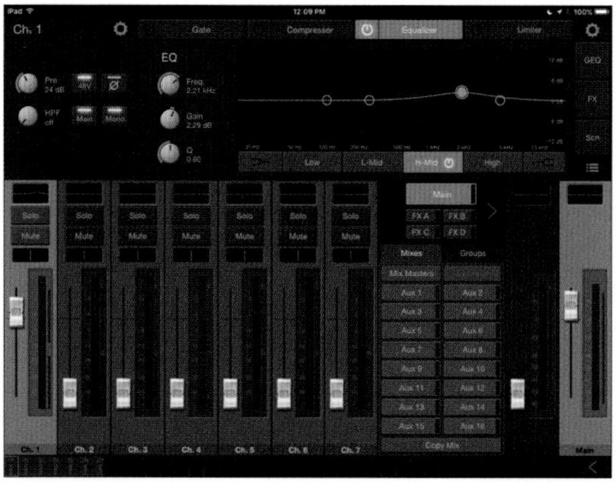

Figure 7.6 The UC Control from PreSonus, shown here running on an iPad, provides control for their interfaces and mixers.

These remote control options can also allow artists to adjust their headphone levels directly, and possibly even create their own mix, without needing to ask the engineer to make the adjustments for them.

Auto-Level

While not particularly common (although I wish it were), some interfaces provide an automated level-setting function. To use this, you can simply turn on the auto-level function and then play the loudest signal you anticipate recording. The interface senses the level and adjusts the input sensitivity to accommodate the loudest signal—but also reserves some additional headroom, just in case. The process ends either after a set period of time or after you turn off the auto-level function (Fig. 7.7).

Figure 7.7 The Zoom UAC-8's Auto function sets level automatically based on the loudest signal it receives.

There are two problems to watch out for when using automatic level settings:

- Sometimes musicians will *think* they've set the auto-level while playing or singing at their maximum level, but in the heat of the moment in a record pass, they end up even louder. Auto-level functionality doesn't work in real time, only prior to recording.
- If the musician doesn't hit the maximum level they preprogrammed, you'll end up losing potential headroom. Fortunately, the audio quality of most interfaces is such that even if a recording peaks well below the maximum available headroom, you'll have no noticeable increase in noise. In fact, many engineers maintain that you will obtain better sound by leaving several dB of headroom rather than always running the level up to maximum.

Digitally Controlled Mic Preamps

Most interface mic preamps have gain knobs for manual level control. As a result, if the interface offers presets or other software control, then the mic preamp levels may not match the knob settings. If your mic preamps have digital gain control, then these settings can be recalled along with the other audio interface parameters (Fig. 7.8). Often these devices will have continuously variable gain knobs that rotate freely in either direction rather than knobs with fixed positions.

Figure 7.8 The PreSonus Studio 192 interface has digitally controlled mic preamps. This economizes on front panel space because left/right buttons choose the channel; there's a single gain control, and you can also enable +48V for each channel instead of needing to do it by groups.

Portability

While there have long been audio interfaces for smartphones and tablets, as interest in mobile applications increases, mobile interfaces are becoming smaller, more affordable, and of higher quality. Some of these are downright ingenious in how much capability they squeeze into a small package—sometimes even including MIDI in and out (Fig. 7.9). If you're a musician on the go, don't ignore audio interfaces just because they're compact.

Figure 7.9 IK Multimedia's iRig Pro has a combi jack for 1/4-inch or XLR plugs, +48V phantom power for condenser mics, and MIDI in and out on 1/8-inch stereo connectors. (Cables for interfacing these MIDI jacks to standard 5-pin DIN connectors come with the package.)

Bundled Software

Many interfaces increase their appeal by bundling additional software. These bundles may include light versions of recording software, plug-ins, loops, sound effects, discount coupons for various related products, and other goodies intended to add extra value.

IEC Cable vs. "Wall Wart"

Plugging a standard IEC AC cord into the back of your interface feels more solid and reliable than using the flimsy wires and small, sometimes proprietary plugs that AC adapters provide. However, expect to pay a bit more for a device with an IEC connection (Fig. 7.10).

IEC: International Electrotechnical Commission, a standards and conformity assessment body for electrotechnology.

Figure 7.10 The receptacle for an IEC AC cord holds the cord securely.

Ground Post

Your interface may have a ground post or simply a screw that connects to the interface's chassis (Fig. 7.11). This can be useful for combating certain kinds of hum or electrical noise.

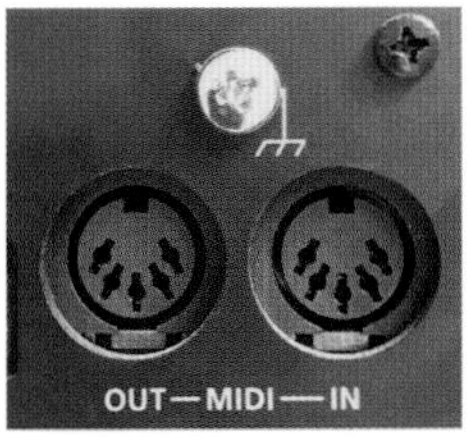

Figure 7.11 A ground post or screw, indicated on this Roland Octa-Capture interface by the little ground symbol above the MIDI jacks, may help reduce or eliminate hum in some situations.

Sometimes if hum is present, connecting a ground lead from the computer's metal chassis (or a patch bay, if present) to the ground post or screw will minimize the hum.

Ergonomics and Build Quality

The ergonomics and build of an interface can be significant factors. Are the controls packed together so tightly they're hard to adjust? For tabletop operation, can you tilt the unit back a bit? If you plan to use the interface for mobile applications, will it survive a fall? Do the knobs wiggle back and forth when you move them? (This can indicate that they're secured to a circuit board, but not to the front panel—which can place more stress on the circuit board over time.) Do the knobs and switches feel substantial?

As with so many other aspects of choosing an audio interface, determining which characteristics matter to you is subjective. For example, I do lots of presentations and carry an interface around with me; as such, I chose a compact interface with a solid metal box that could survive TSA searches and falls.

Front Panel vs. Rear Panel Jacks

Most audio interfaces put the instrument jacks on the front panel, to allow for easy plugging and unplugging of guitars and basses. Some put the majority of their connections on the front, turning the interface into somewhat of a mini-patch bay. Others have parallel connections on the front and back.

Evaluate your situation: If you'll mostly be plugging in devices permanently to the interface (e.g., synthesizer outputs), then having the connections on the back will help keep a rat's nest of wires out of the way. This is particularly useful if you've expanded to multiple inputs specifically for this purpose. However if your interface will be serving double-duty as a patch bay, then it may be helpful to have all the connectors on the front.

Key Takeaways

- Quality input level metering on the front panel can be helpful when setting levels directly on an interface.
- Auto-level can be a useful feature when used properly.

- IEC AC cords are more convenient than AC adapters, but an AC adapter may not be a deal-breaker.
- Whether it matters if most of the jacks are on the front or back depends on your studio setup. It's convenient to have guitar inputs on the front panel, but this may not matter if you're using something like a Kemper Profiling Amp or Line 6 Helix, where you'll probably plug into a pair of line inputs.
- If you're always re-patching different mics, amps, instruments, etc. into your interface, it's convenient to have as many jacks as possible on the front panel. If your setup is more permanent, rear-panel connections will help reduce clutter.

Chapter 8

Cascading/Aggregating Interfaces

If you already have a go-to interface that you like, that's great. But what do you do when your projects expand and you find that you need more inputs? Fortunately, you can configure your recording software to address more than one interface—but certain rules apply. In this chapter, we'll take a look at some available options for aggregating your devices.

Windows

ASIO was not designed to handle multiple audio interfaces. Some manufacturers such as RME, Roland, and PreSonus circumvent this with custom drivers. These drivers allow certain interfaces to be used simultaneously and treated as one big interface instead of multiple smaller ones. This can work quite well within the constraints that the manufacturer imposes—the main one being that the process works only with that manufacturer's interfaces, or only with their interfaces that use certain protocols (e.g., FireWire or Thunderbolt).

However, Windows' native drivers like WDM/KS, WASAPI, and WaveRT can indeed handle multiple interfaces, provided that the interfaces are compatible with those drivers. When you select an input or output in a track, all the interfaces that support those drivers will show their available inputs or outputs respectively in your application (Fig. 8.1).

Figure 8.1 Acoustica Mixcraft, when set to Windows drivers instead of ASIO, shows all available hardware inputs when selecting a track input for recording. You can also select a default driver. In this example, the Line 6 Helix audio interface is being selected as the track input.

Apple Mac

Apple's Core Audio was designed from the ground up to allow aggregating interfaces. To configure this, open Audio MIDI Setup, and then choose the Audio Devices window. Next, click the little + sign in the lower left corner and choose `Create Aggregate Device`; an Aggregate Device option appears, and you'll see a list of your available I/O, based on your currently connected devices.

Place a check next to each of the interfaces you want to aggregate, and then choose the desired clock source—most of the time whatever's selected when you aggregate the interfaces will work (if not, try a different clock source). If there's no hardware sync between the interfaces (i.e., there's no word clock connection), check the Drift Correction boxes on the right side (Fig. 8.2). Now all input and output options will be available in your host program.

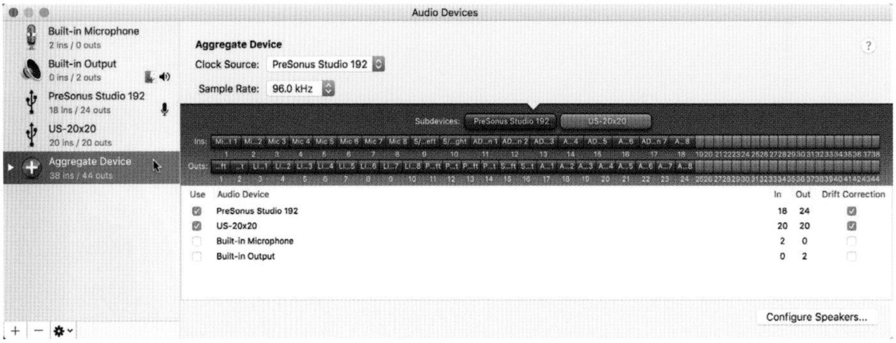

Figure 8.2 Using Core Audio on the Mac with a PreSonus Studio 192 and TASCAM US-20 × 20 aggregated to create a single 38 input/48 output interface for 96 kHz operation.

If you encounter any problems, go to the Audio MIDI Setup's Help and search on Aggregation. Choose Combining Audio Devices, and follow the directions.

Expanding Interfaces via the ADAT Port

Because I do quite a bit of solo work, I usually need only a few inputs at a time. However, I prefer to have interfaces with enough inputs that I don't need to re-patch, allowing me to leave instruments set up for inspiration at a moment's notice. Because combining different interfaces via ASIO can be problematic, it's fortunate there's an alternative: using ADAT-compatible inputs and outputs.

Setup

My main interface has eight mic/line inputs, which is not enough to handle the multiple outputs available on the various hardware synthesizers I use, not to mention guitar, bass, microphones, etc. However, this interface also has two ADAT inputs. Many other interfaces can send their output signals via ADAT output, making them suitable as expanders for this interface.

For example, the TASCAM US-20 × 20 interface has an ADAT optical output and also provides the option to switch it into Mic Pre mode. This means it no longer needs to connect to the computer via USB, and the eight mic inputs will feed both the analog and ADAT outputs simultaneously (Fig. 8.3). Since I can control gain directly from the US-20 × 20's front panel controls, it's not necessary to load its mixer application.

Interfaces that don't have a mic pre mode are still suitable as long as you can route their preamps to an ADAT-compatible output.

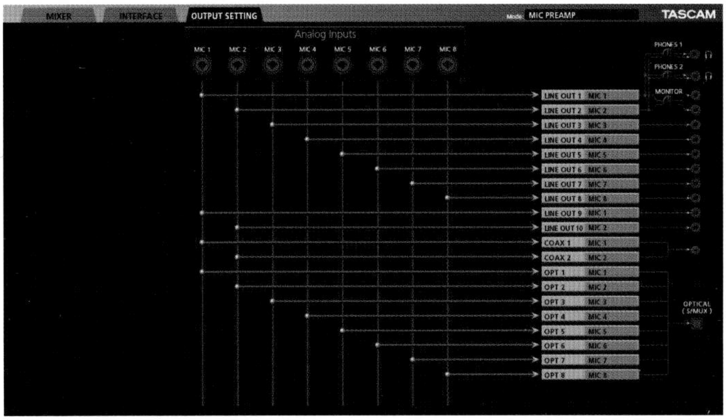

Figure 8.3 In Mic Pre mode, the US-20 × 20's mic preamp outputs go to both analog and ADAT digital outputs.

Other units like the MOTU 8pre, Focusrite OctoPre, PreSonus DP88, and more are also designed specifically to be mic preamps with ADAT and/or analog outputs for expanding existing audio interfaces. Note that the signals carried over ADAT are digital, so you'll get none of the audio degradation that could occur with an analog sub-mixer—the signal appearing at your recording software is no different from the signal coming from the expansion device's mic preamps.

Some units allow for analog and/or digital expansion. The analog expansion is typically provided through DB-25 connectors, which have 25 pins (Fig. 8.4). These can wire directly to patch bays in larger studios via a multi-conductor cable, which makes for a neater setup than having separate cables connected to multiple inputs.

Figure 8.4 In addition to including ADAT-compatible inputs and outputs, the PreSonus DigiMax DP88 makes the line inputs, mic inputs, direct outs, and digital-to-analog converter outputs available on separate DB-25 connectors.

Interconnecting an Audio Interface and Expansion Unit

The process for interconnecting two units is pretty painless. Use the following steps:

1. If the unit intended to expand your interface has a special mode for sending its inputs to the ADAT output, select it (e.g., the TASCAM US-20 × 20's Mic Pre mode).
2. Patch a TOSLINK optical cable from the expander unit's ADAT out to your main interface's ADAT in.
3. Open up your main interface's setup or mixer application, and (important!) choose ADAT as the sync (or clock) source (Fig. 8.5).

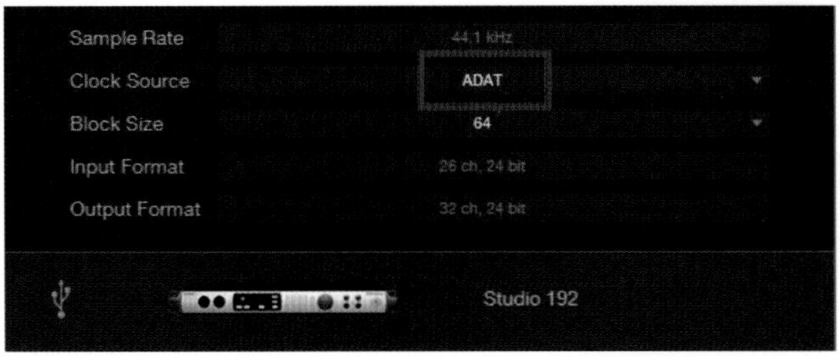

Figure 8.5 Choosing ADAT as a clock source for the Studio 192 interface allows the interface to sync properly to the expansion unit.

The output signals from the expander will now be available through your main interface. You may need to enable these in a preferences menu or I/O setup dialog box as new potential inputs or otherwise assign the expander's outputs to your recording software's inputs (Fig. 8.6).

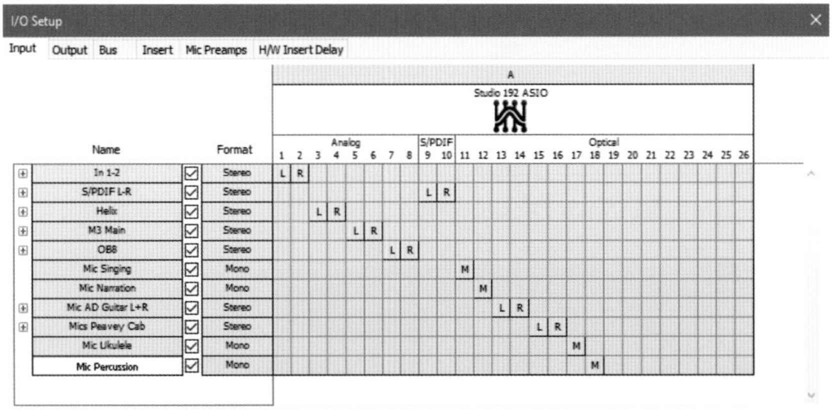

Figure 8.6 In Pro Tools, the eight channels from the DP88 expander appear as optical inputs 11-18, which are linked to specific inputs within Pro Tools. For example, if you choose "Mic Narration" as an input for a channel in Pro Tools, it will record the expander signal coming in on optical input 12.

Although that's pretty much all you need to do, here are a few fine points to consider:

- A single ADAT optical connection carries eight channels at 44.1 and 48 kHz sampling rates. At higher sampling rates, when using the S/MUX modes described in Chapter 5, the channel count goes down (four channels at 96 kHz or two channels at 192 kHz). Note that not all devices can send and/or receive ADAT signals at higher sample rates.
- Adding a second ADAT optical connection doubles the potential number of channels.
- You won't be able to set a sample rate higher than 44.1 or 48 kHz if the main interface syncs to the ADAT clock and the expander module doesn't support S/MUX sample rates.
- The expansion interface doesn't have to be next to your main interface; the distance is limited only by the optical cable. The ADAT spec assumes a maximum length of about 16 feet, but with high-quality optical cables, this can extend much further. So for example, you could place the expander next to a keyboard rack and avoid running a bunch of cables from your keyboards to your main interface.

Key Takeaways

- If you use Windows and think you'll need to add more inputs, choose an interface you can expand via an ADAT port. Aggregating through the Windows native drivers generally leads to more latency and less efficient operation.
- Aggregating on the Mac is really easy. You can add more inputs simply by buying another interface or using ADAT-based expansion.
- When expanding via an ADAT port, sync the main audio interface to the clock signal appearing at the ADAT input that is connected to the expansion unit's ADAT output.

Appendix

Obtaining the Benefits of High Sample Rates in Lower Sample Rate Projects

The controversy about whether audio recorded at 96 kHz sounds better when played back at 96 kHz than when played back at 44.1 kHz has never really been resolved. However, the sample rate can affect sounds generated "in the box," such as a virtual instrument plug-in that synthesizes a sound or an amp simulator plug-in that creates a distortion effect. Any improvement heard with high sample rates comes from eliminating *foldover distortion*, also known as *aliasing*.

Theory time! Let's start with a simplistic example: A digital system can accurately represent audio only at frequencies lower than half the sampling rate (called the Nyquist limit—e.g., 22.05 kHz in a 44.1 kHz project). If an algorithm within a plug-in generates harmonic content above this limit—for example, at 40 kHz—then you won't hear this content; however, you will hear the aliasing created when this 40 kHz tone "folds down" below the clock frequency (to 4.1 kHz, in this case: the difference between the 44.1 kHz clock frequency and the 40 kHz tone). Aliasing can thus appear within the audible range, but it is harmonically unrelated to the original signal and generally sounds pretty ugly.

For more information on aliasing, refer to the *Sound On Sound* article on the causes of aliasing, available here: http://bit.ly/sos-alias.

Foldover distortion can happen with synthesized waveforms that are rich in harmonics, like pulse waves with sharp rise and fall times. (Amp sims can also be problematic; although the harmonics may be weak, when you apply 60 dB of gain to create overdrive or distortion, they can be strong enough to cause audible aliasing.)

Not all plug-ins will exhibit these problems, for a variety of reasons:

- The audio may not be rich enough in harmonics (or distorted enough) to cause audible aliasing.
- The plug-in itself may oversample internally, meaning that within the plug-in, the sample rate is higher than that of the recording. In this case, any foldover distortion will occur outside the audible range.
- The project sample rate may be high enough to provide the same kind of environment as oversampling.
- The plug-in designers may have included appropriate anti-alias filtering in the plug-in.

Many modern virtual instruments and amp sims provide oversampling, and recording software can handle very high sample rates, so you'd think that might be the end of it. Unfortunately, other limitations can apply with oversampling and higher project sample rates.

- Recording an entire project at a high sample rate stresses your computer more, reduces the number of audio channels you can stream, and probably won't allow you to run as many plug-ins.
- Oversampling requires more CPU power, so even if all your instruments are oversampling internally, you may not be able to use as many instances of them.
- Although some instruments may perform 2× oversampling, that still might not be sufficient to eliminate aliasing on harmonically rich sources.

Furthermore, the sound quality of plug-ins that oversample will be influenced by the quality of the sample-rate conversion algorithms they use. These can be a source of signal distortion themselves.

Fortunately, a couple of solutions are available.

Solution 1. One solution is simply to temporarily change the sample rate within the recording software from 44.1 or 48 kHz to something higher, like 96 kHz. If the audio interface can adjust itself to the new sample rate, then all you need to do is render the processing on any tracks that would benefit from the higher sample rate. This converts them into audio at the higher sample rate, so there won't be any audible foldover distortion. After rendering all the sounds that benefit from the higher sample rate, you can then change the sample rate back down to your original project rate of 44.1 or 48 kHz.

If a project contains only virtual instruments and no audio, you can generally freely change the sample rate as you work.

You will not lose the benefits of upsampling when you later convert the sample rate back down to 44.1 kHz, because rendering at the higher sample rate will have eliminated any foldover distortion in the audible range—and 44.1 kHz has no problem playing back sounds in the audible range.

However, note that upsampling can't fix audio already affected by aliasing distortion; upsampling audio that already contains foldover distortion will simply reproduce the existing distortion. This technique applies only to audio *created* in the computer at the higher sample rate. Similarly, it's unlikely that upsampling sound recorded via a computer's audio interface will yield any benefits, because the audio interface itself will have already band-limited the signal's frequency range to eliminate harmonics that interfere with the clock frequency.

Solution 2. In some situations you will not be able to change the song's sample rate easily. Here's another method you can use if Solution 1 won't work. Let's assume that you are working on a 44.1 kHz project with a virtual instrument and that the virtual instrument's MIDI track has been finalized but not rendered to audio yet.

Use the following steps to upsample the virtual instrument:

1. Save the virtual instrument preset so you can call it up in Step 5.
2. Export the MIDI clip driving the instrument.
3. Close the existing project.
4. Create a new project at a higher sample rate, like 88.2 or 96 kHz.
5. Insert the virtual instrument you used previously, and then load its preset.
6. Import the MIDI clip and assign it to the Instrument track.

> Be sure to import the original tempo with the MIDI clip, or set the tempo in your new session to match that of the original session.

7. Render the instrument track starting at the beginning of the song.
8. Export the rendered audio; then close the project.
9. Open the original project with the 44.1 kHz sample rate.
10. Import the rendered audio; your recording software will convert the sample rate back to 44.1 kHz. Use this as the instrument audio instead of rendering the instrument from the 44.1 kHz project.

Note that some DAWs may allow you to bounce a track from your original project at a higher sample rate. This will produce the same results for the bounced file as the above workflow, allowing you to avoid the trip to a high-sample rate project. After bouncing the virtual instrument track, simply import the bounced file using sample rate conversion to use it in your 44.1 kHz project.

Final Thoughts

At times you may not *want* the difference caused by upsampling. When I did an upsampling demo at a seminar with a particular synthesizer, most attendees preferred the sound with the aliasing because the upsampled sound was brighter than what they expected. However when I upsampled an amp sim and with a different synth, the consensus was that the upsampled versions sounded much better. Regardless, the point is that upsampling gives you a choice—use the high-resolution instrument if you like that sound better, or leave it as it is with any distortion caused at a lower sample rate. After all, distortion isn't necessarily always horrible—think of how many guitar players wouldn't have a career without it.

Although upsampling isn't a panacea, don't dismiss it either. With many synths, upsampling will make no audible difference—but with others, you'll hear more clarity. As always, use your ears to decide what sounds best for your music.

About the Author

Musician/author Craig Anderton is an internationally recognized authority on music and technology. His onstage career spans from the 60s with the group Mandrake, through the early 2000s with electronic groups Air Liquide and Rei$$dorf Force, to the "power duo" EV2 with Public Enemy's Brian Hardgroove, and EDM-oriented solo performances.

He has played on, produced, or mastered over 20 major label recordings, did pop music session work in New York in the 1970s on guitar and keyboards, played Carnegie Hall, and more recently, has mastered well over a hundred tracks for various artists.

In the mid-80s, Craig co-founded *Electronic Musician* magazine. As an author, he's written over 26 books on musical electronics and over a thousand articles for magazines like *Keyboard, Sound on Sound, Rolling Stone, Pro Sound News, Guitar Player, Mix,* and several European publications.

Craig has lectured on technology and the arts (in 10 countries, 38 U.S. states, and three languages), and done sound design work for companies like Alesis, Gibson, Peavey, PreSonus, Roland, and Steinberg.

Please check out some of his music at youtube.com/thecraiganderton, visit his web site at craiganderton.com, and follow him on twitter @craig_anderton.